Encounter World Regional Geography

INTERACTIVE EXPLORATIONS OF EARTH

Using Google Earth™

JESS C. PORTER

Prentice Hall

Boston Columbus Indianapolis New York San Francisco Upper Saddle River
Amsterdam Cape Town Dubai London Madrid Milan Munich Paris Montréal Toronto
Delhi Mexico City São Paulo Sydney Hong Kong Seoul Singapore Taipei Tokyo

Geography Editor: Christian Botting
Managing Editor, Geosciences: Gina M. Cheselka
Project Manager: Wendy A. Perez
Marketing Manager: Maureen McLaughlin
Media Producer: Ziki Dekel
Editorial Assistant: Christina Ferraro
Marketing Assistant: Nicola Houston
Editorial Director, Geosciences: Frank Ruggirello
Cover Designer: Richard Whitaker
Operations Specialist: Maura Zaldivar

Cover Photo Credit: NASA
(Left) Access Code Card Screen Capture Credits: ©2009 Google, ©2010 Geoeye, ©2010 Europe Technologies
(Right) Access Code Card Screen Capture Credits: ©2010 TerraMetrics, ©2010 Geocentre Consulting, ©2010 DigitalGlobe, ©2010 Geoeye
(Left) Back Cover Screen Capture Credits: ©2010 Europa Technologies, ©2010 Geoeye, ©2010 ZEN-RIN, ©2010 DigitalGlobe
(Right) Back Cover Screen Capture Credits: ©2010 TeleAtlas, Data SIO, NOAA, U.S. Navy, NGA, GEBCO, US Dept. of State Geographer, ©2010 Europa Technologies

Pearson Prentice Hall
Pearson Education, Inc.
Upper Saddle River, New Jersey 07458

Printed in the United States of America
10 9 8 7 6 5 4 3 2 1

ISBN-13: 978-0-321-68175-1
ISBN-10: 0-321-68175-4

Prentice Hall
is an imprint of

www.mygeoscienceplace.com
www.pearsonhighered.com

Contents

Preface

Welcome to *Encounter World Regional Geography*! This workbook will immerse you in interactive explorations of the world's immense geographic diversity. Elements of physical geography, human geography, and geospatial techniques come together to give you a better understanding of our world regions' idiosyncrasies and areas of common ground. This is accomplished by applying the power of the Google Earth™ program to zoom-in and around features and landscapes ranging from street-corner scenes to regional ecosystems. Within Google Earth™, we will utilize associated tools and layers such as photographs, satellite imagery, and historic maps. We will springboard into related websites that help us understand the patterns and processes taking place on Earth and among its diverse peoples.

As you work through the exercises contained in this book, you will feel your "big picture" understanding of the world come into focus. Not only will your knowledge of the themes of geography grow, but also your ability to apply these themes to your interpretation and understanding of life on Earth. While these exercises are designed to educate thematically and topically, they are also designed to be fun. We encourage you to take these exercises beyond the parameters laid out in the multiple choice and short answer question segments of the workbook. If something piques your curiosity, dig-in deeper. Look for the answers, but also look for more questions to ask. It is our hope that you will apply your enhanced spatial understanding of the world and its regions beyond your studies associated with this workbook. You will find that improving your spatial thinking skills is something that can benefit every aspect of your life.

This workbook is organized with two introductory chapters. The first provides you with a Google Earth™ primer as you learn to navigate and use the software's system of layers to display spatial information. The second chapter addresses some key concepts in geography that are essential for you to get the most out of this workbook. Location, scale, and place are discussed and the basics of interpreting remotely sensed images are introduced. This will help you better understand and interpret what you are seeing in the Google Earth™ environment. The workbook then examines 12 commonly identified world regions, beginning with North America.

Each of these chapters is organized around five themes of geography: Environment; Population; Culture; Geopolitics; and Economy and Development. Within each thematic exploration, you will have the opportunity to answer multiple-choice and short essay questions. These questions have been designed with an emphasis on high-level assessment skills that will encourage you to interpret and appraise the imagery and information that you uncover. Additionally, the You Map It! activity that concludes each chapter provides you with the chance to employ some of the same mapping techniques and tools that have been utilized to create the explorations contained in this workbook. Good luck and have fun!

Jess C. Porter
jcporter@ualr.edu

Encounter World Regional Geography

Name: ______________________________

Date: ______________________________

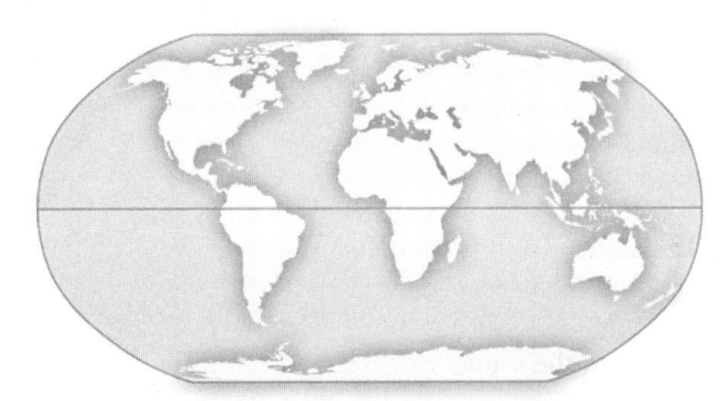

Chapter 1: Introduction to Google Earth™

Verify that your computer has Google Earth™ software installed. If you do not, then go to http://earth.google.com and download the most recent version of the free Google Earth™ software. Your computer will need a high-speed internet connection in order to utilize the software effectively.

Download EncounterWRG_ch01_Intro.kmz from ***www.mygeoscienceplace.com*** *and open in Google Earth™.*

Exploration 1.1: GETTING STARTED WITH NAVIGATION

The very best way to learn Google Earth™ is to dive in and explore your world. The software is intuitive to use and not complicated. However, you will also find it useful to learn some of the basics of navigation as demonstrated in the User Guide. Go to http://earth.google.com/support and then click the link for User Guide. Do not forget about this useful resource. If you are ever confused or need help with using the software, this will be your best bet to find assistance.

In the User Guide, under the "Getting Started" heading, there is a link titled "Getting around." Open this link and study the "Getting to Know Google Earth™" diagram and associated information. Be sure to click the links associated with the tools to find out about some of the capabilities of Google Earth™.

Exploration 1.1: MULTIPLE CHOICE

1. If you needed to find a specific place on Earth, you would use the

A. Places panel.
B. Layers panel.
C. Overview map.
D. Finder tool.
E. Search panel.

2. The on-screen navigation controls for Google Earth™ are located on what part of the Google Earth™ window?

A. bottom right
B. top right
C. center
D. bottom left
E. top left

3. Which of the following is *not* a capability of Google Earth™?

A. You can display sunlight across the landscape.
B. You can add polygons, lines, and points (placemarks) to the view.
C. You can view live imagery.
D. You can display historical imagery.
E. You can email views from Google Earth™.

4. Where would you find the information regarding a view's coordinates, elevation, and imagery date?

A. In the Places panel.
B. In the Search panel.
C. In the Layers panel.
D. In the 3D Viewer.
E. In the Overview map.

Now click the "Navigating in Google Earth™" link, located under the "Getting around" heading. Play the short video to learn how to navigate the globe in Google Earth™. Scroll down the page and review other tips for navigating in Google Earth™.

5. If you wanted to look around from one vantage point, as if you were turning your head, you would use the

A. on-screen look joystick.
B. arrows on your keyboard.
C. on-screen move joystick.
D. wheel on your mouse.
E. on-screen zoom slide.

Click the "Five Cool, Easy Things You Can Do in Google Earth™" link, located under the "Getting around" heading. Enter the location of your university in the Search panel and then view the image of your university.

Exploration 1.1: SHORT ESSAY

1. Seeing your university from an aerial perspective, describe some insight that you have gained. Did you find out there were buildings you did not know existed? Maybe there is a parking lot that is closer to your dorm than the one you currently use? Feel free to be creative in your response.

____
____
____

2. Navigate using both the on-screen controls such as the zoom slide and move joystick; then navigate using the keyboard and/or mouse. Try integrating the two approaches. Which method(s) do you prefer and why?

____
____
____

Exploration 1.2: LAYERS

Google Earth's™ functionality is magnified by the ability to add layers of data and information to the 3D Viewer. Beyond the visual representation of the Earth's surface we can add any number of features such as the road networks, place names, boundaries, photographs, business locations, current weather conditions, and three dimensional buildings to name just a few. A problem that faces many users of Google Earth™, however, is that their viewers become cluttered with too much information. In general, it is best practice to simply display only the layers you need at any given time. *Borders and Labels* and *Terrain* are the two layers that you will find most useful without cluttering the view.

Begin by going to the Layer pane and turning off all of the layers. This can be done with a single click by un-checking the box next to *Primary Database.*

Open the *Intro to Google Earth*™ folder that is loaded in the *Temporary Places* folder in the Places pane. You can right-click on the Intro to Google Earth™ folder and you will see an option to "Save to My Places." If you select this, the folder will be moved to your *My Places* folder and you will not have to reload it next time you start Google Earth™. If you don't do this now, you will be given the option when you exit Google Earth™ at the end of your session.

By clicking the plus sign next to the *Intro to Google Earth*™ folder you will expand the folder and see what's inside. You will see a series of five folders. Expand the *1.1 Layers* folder. Expand it and then double-click the *Albuquerque, NM* placemark. Now check the boxes beside the dozen or so layers and/or folders in the *Primary Database.* Wait a few moments and you will get a taste of how cluttered Google Earth™ can become. Now turn off all of the layers/folders except the folder called *Borders and Labels* and the layer called *Terrain.* Depending on the type of question you seek to answer, you may want to activate one or several of the folders or layers at a time.

For example, let's start with the *Panoramio Photos* layer. Turn this layer of information on by clicking the box next to *Panoramio Photos* in the *Primary Database.* You will see a number of small blue squares appear on the screen. Begin clicking these blue squares and you will see that these are photographs of features that are correlated with their location on the ground. Click on approximately ten of these boxes contained in the current view. Do not adjust the view by panning or zooming.

Turn off the *Panoramio Photos* layer and turn on the *Roads* layer. You will see that the major highways of the community are illuminated and labeled. This *Roads* layer can be very helpful when you are attempting to analyze patterns on the landscape as transportation is closely related with many types of development. Now use the zoom-slider or the scroll wheel on your mouse to slowly zoom in. As you zoom in you will notice that more and more details of the road network will emerge until the smallest residential streets appear and are labeled. Zoom out and these features and labels disappear.

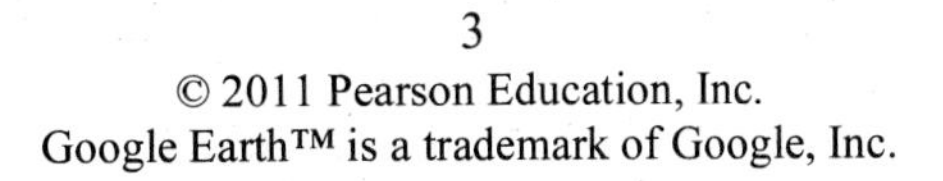

Exploration 1.2: MULTIPLE CHOICE

1. The name of the street that is immediately north of the Albuquerque, NM placemark is

A. Indian School Rd.
B. Interstate 40
C. State Highway 47
D. Richmond Dr.
E. Euclid Ave.

Turn off the *Roads* layer and turn on the *Terrain* layer. Double-click the *downtown* placemark in the Places pane to zoom to a view of downtown Albuquerque. Perhaps you can get a sense that this is the central business district by the relatively larger buildings. Maybe you noticed the shadows in the image suggesting some of these buildings had considerable vertical development. It would be helpful if we had more information that could be added to the map to help us understand the physical realities of downtown. Turn on the *3D Buildings* folder and wait a moment for the data to load. You can click on individual 3D models that have been built to gain additional information. Now you have a much better feel for what the downtown area really looks like.

This added information is what geographic information systems (GIS) are all about. We can take layers such as a road network and do analysis in relation to the built environment. Throughout this text you will be capitalizing on the ability of Google Earth™ to display and overlay different types of data and information on the surface of the digital globe in order to increase your understandings of the world and its regions.

2. What is the name of the tallest building in downtown Albuquerque?

A. New Mexico Bank and Trust
B. Albuquerque Convention Center
C. Commerce Bank of Albuquerque
D. Hyatt Regency / Bank of Albuquerque
E. Albuquerque Petroleum Building

Turn off the *3D Buildings* folder and turn on the *Street View* layer. After a moment you will see camera icons emerge on the streets of the city. This indicates that street-view imagery is available for that segment of road. Street-view imagery is increasingly available in urban locations around the world. This tool can really do a great job of providing you with a sense of place of a particular location as you can simulate driving down a road by clicking on successive camera icons. From the perspective of the *downtown* placemark, select a camera icon on Central Avenue that is near the yellow push-pin for *downtown.* Click the camera icon, the click the link to "Show full screen." Work your way down Central Avenue to the 312-318 SW address range. There is a bar and grill with a unique sign.

3. The bar and grill that is located in the vicinity of 312-318 Central Avenue SW has a unique sign that identifies it as the

A. Library Bar and Grill.
B. Eagle and Child Bar and Grill.
C. Ritz Bar and Grill.
D. Long Bar and Grill.
E. Celtic Bar and Grill.

Click "Exit Photo" in the upper-right corner when you have completed your street-viewing. Turn off the *Street View* layer. Turn on the *2009 drought images* layer to see an illustration of the application of areal data to Google Earth™. You will see that a time slider has appeared in the upper-left of your screen. It indicates that the data shown represents conditions on January 5, 2009. Here it would be helpful to know what states are affected, so let's turn on the *Borders and Labels* folder. Sometimes it's advantageous to make a data layer somewhat transparent. At the bottom of the Places pane you will see a slider that will allow you to adjust the transparency of an overlay. Practice manipulating the transparency before returning the slider to the maximum coverage (slid to the right). Then start the animation by clicking the play button in the time-slider. It will likely take several repeats before all the images load.

4. The USDA drought imagery illustrates that drought conditions were consistently most severe from January 2009 to August 2009 in

A. Florida.
B. the Great Lakes region.
C. an area from Maine to New York.
D. south Texas.
E. the Dakotas.

Turn off the *2009 drought images* layer. In the *Primary Database*, expand the *Weather* folder. Turn on the *Clouds* layer and explore the coverage of this data layer and then turn on the *Radar* layer. You will see that these layers do not have the same coverage. While cloud imagery is available globally, the radar data is only available for select locations like North America and Western Europe. This data is one of the rare occasions in Google Earth™ where you are viewing near-live data. When you have completed your exploration of clouds and precipitation, turn off and collapse the *Weather* folder.

There are dozens of additional datasets, multimedia files, and links to outside resources in Google Earth™. For a taste, expand the *Gallery* folder. Turn on the *NASA* folder and then navigate to southern Ethiopia. Explore the NASA content that is available for this location.

5. What statement best describes the NASA data available associated with the NASA icon positioned over southern Ethiopia?

A. Volcanic activity that is ongoing in the region.
B. Effects of Saharan dust storms on Ethiopia's cities.
C. Imagery of vegetation deviation from normal.
D. Pollution from industrial fires.
E. Copper mining impacts on the landscape.

Exploration 1.2: SHORT ESSAY

1. Based solely on what you can ascertain from your sampling of Panoramio photographs, write a short paragraph that describes the community of Albuquerque, New Mexico.

___

___

___

2. Describe the clouds and precipitation patterns over North America using the Radar data from Google Earth™. Is there a strong correlation between cloud cover and precipitation? Can you identify a storm system(s) (area(s) of low pressure)? Where are these located? Has or will this storm system(s) affect you at your current location? Document the date and time you viewed these features in your response.

____

____

____

Exploration 1.3: ENVIRONMENT AND POPULATION

Let's continue to build our familiarity with Google Earth™ by exploring some locations that introduce the thematic framework of this workbook. Environment, population, culture, geopolitics, and economy and development represent the five themes around which the explorations are built. As you complete the explorations, you will find that these five themes overlap and intertwine with regularity.

The environment theme represents the dynamic habitat of humanity and the stage upon which many of our anthropogenic-oriented geographic phenomena play out and interact. In this text we will explore unique features and landscapes of physical geography as well as the unparalleled capacity of people to alter and impact them.

Demographic and settlement characteristics of the human population will be explored in our second theme. From population density to population growth rates and life expectancies, the visualizations of Google Earth™ shed light on the patterns of people.

Google Earth™ is a terrific tool to visualize Earth's complex physical landscapes. The three dimensional capacity of the program allows the user to get a grasp on the variations in relief that a paper-map is not able to provide. In the *Primary Database*, verify that all layers are turned off except for *Terrain*. Open the Google Earth Options dialog box by selecting *Tools*, and then *Options* from the menu bar. On the *3D* tab, verify that elevation is displayed using *Meters, Kilometers*. Open the 1.3 ENVIRONMENT AND POPULATION folder and double-click the *Grand Canyon* placemark. You will zoom to a view of this unique natural feature. As you move the cursor across the screen you will see the values for elevation that are displayed at the bottom of the screen change.

Exploration 1.3: MULTIPLE CHOICE

1. What is the approximate elevation value of the Colorado River as displayed in the Grand Canyon placemark?

   A. 475 meters above sea level
   B. 1145 meters above sea level
   C. 1550 meters above sea level
   D. 2400 meters above sea level
   E. 6335 meters above sea level

Turn on the *Deepwater Horizon* placemark and then double-click it to fly to the location. This is the site of 2010's oil spill that occurred after a deep-sea drilling rig experienced a total failure. Let's use the measure tool (the button with a ruler) to determine the distance between the site of the accident and the nearest land. Click the ruler tool and then position the cursor over the Deepwater Horizon placemark. Click once and the move the cursor over the nearest land (the end of the Mississippi River delta) and click again. Be sure you have changed the units to kilometers.

2. What is the approximate straight-line distance between the Deepwater Horizon site and the nearest land?

A. 20 kilometers
B. 50 kilometers
C. 80 kilometers
D. 120 kilometers
E. 175 kilometers

An important part of monitoring the spill in the days, weeks, and months after the event is the application of satellite imagery. Turn on the *2010-05-17 MODIS* folder and zoom to an appropriate level to view the extent of the spill.

3. At the time the 2010-05-17 MODIS image was captured, the oil spills extended farthest from the Deepwater Horizon site in the

A. western direction.
B. northern direction.
C. eastern and northern directions.
D. northern and western directions.
E. southern and eastern directions.

Now turn on and double-click the *New Orleans, LA and suburbs flood zones* layer. This map has been used as a guide for residents of New Orleans who have decided to rebuild following Hurricane Katrina. Notice the slider at the bottom of the Places pane. You can use this to make the active layer (the one that is highlighted in the Places pane) more or less transparent. Experiment with the slider now.

4. The area of New Orleans that is generally lowest in elevation is

A. the Algiers neighborhood, south of the Mississippi River.
B. the area south and east of the Lakefront Airport.
C. downtown.
D. the area due east of Louis Armstrong International Airport.
E. the Elmwood neighborhood, north of the Mississippi River.

Let's take a look at a sample of population data displayed in Google Earth™. Double-click and turn on the *life expectancy at birth – 2008*, folder. The height and color of the country polygons correlate to the life expectancy values. The countries that are more extruded (taller/higher) have values that are higher. Colors are ramped to illustrate locations at the top and the bottom of the scale. You can click on the flag of a country to see its vale, as well as the highest and lowest value in the data set.

5. Which of the following countries has the lowest life expectancy at birth in the 2008 data set?

A. Afghanistan
B. Japan
C. Laos
D. United States
E. Zambia

Exploration 1.3: SHORT ESSAY

1. Based on what you have seen regarding the Deepwater Horizon oil spill and the rebuilding of New Orleans, identify and describe two ways Google Earth™ can be applied as a beneficial tool in times of crisis.

____
____
____

2. Describe the patterns that you see in the life expectancy layer from a world regions perspective. For example, are there any regions that have generally higher or lower life expectancies? Identify several factors that can contribute to the patterns you have identified.

____
____
____

When you have completed this exploration, turn off the 1.3 ENVIRONMENT AND POPULATION folder.

Exploration 1.4: CULTURE AND GEOPOLITICS

Culture and geopolitics are our next themes. When we speak of culture, we can think of any of the many variables that define a respective "way of life." For example, religion, language, and ethnicity all contribute to defining cultures on a regional or local scale. The interaction between power, territory, and space is a manifestation of culture that is granted its own thematic focus in this text. The ongoing fragmentation and unification of political entities around the world merits this focus.

Begin our sampling of cultural and geopolitical phenomenon by double-clicking and turning on the *USA religious adherents* layer. Counties are enumerated according to the percentage of persons per county that adhere to a religious body. Assess the map for patterns.

Exploration 1.4: MULTIPLE CHOICE

1. Which of the following statements is *not* supported by the USA religious adherents layer?

   A. Utah has one of the highest state rates of religious adherence.
   B. Oregon displays one of the highest levels of religious adherence.
   C. The east and west coasts generally have lower levels of adherence than the interior of the US.
   D. Religious adherence in the Great Plains is generally higher than in the Rocky Mountains.
   E. The map suggests that Illinois is more religious than Indiana.

Some people have a near religious feeling for the popular culture location associated with the *cultural icon* placemark. Double-click this placemark. Study the site and the surrounding area and use clues provided by other layers you can turn on in the *Primary Database*. When you complete the following question, turn off any layers except *Terrain* that you have turned on in the *Primary Database*.

2. The site associated with the cultural icon placemark is

A. Las Vegas.
B. the Mall of America.
C. Disneyland.
D. the state capital of Arizona.
E. the Neverland Valley Ranch.

Within the aforementioned cultural icon site, there is at least one building that could be identified by many people around the world. The built environment can provide us with buildings that reflect both folk and popular culture and the values of a local society in general. The next folder, *unusual buildings*, is a fun collection of interesting structures from around the world. Click on the placemarks in the map or in the Places pane to see pictures of these structures.

3. In what state is found a unique building that suggests fishing is a significant part of the local culture? Remember you can turn on the Borders and Labels layer see the states labeled.

A. California
B. Georgia
C. Maryland
D. New York
E. Wisconsin

You might not think of Google Earth™ as a law enforcement tool, but that is exactly what it was when Swiss police were investigating some farmers they suspected of involvement with illegal drugs. Their Google Earth™ survey of the suspected farmers' lands revealed that the law was being broken. Double-click the *caught* placemark.

4. Identify the statement that best describes the Google Earth™ evidence.

A. The farmers had stockpiled weapons on the site.
B. Large barricades had been constructed along the roads leading to the farmers' compound.
C. The farmers had burned all the native vegetation so they could grow their drug crop without competition for nutrients.
D. The farmers were hiding one type of crop inside another type of crop.
E. The farmers had located themselves in a location with no neighboring farms or houses.

Now double-click and turn on the *Israel's security fence* layer. The barrier is a network of fences, trenches, and concrete walls. When completed, the fence will be more than 700 kilometers long. Supporters of the barricade claim that it will protect Israeli's from terrorism while people opposed to the barrier claim it severely impacts Palestinians ability to move within and out of the West Bank and is a violation of international law. Examine the plan for the fence and compare it with established borders in the region that can be seen by activating the *Borders and Labels* layer.

5. Which of the following statements regarding the security barrier is least accurate?

A. The barrier has deviations in the north that add more territory to Israel compared to the 1949 Armistice Line.
B. The barrier has deviations in the south that add more territory to Israel compared to the 1949 Armistice Line.
C. In some locations, the barrier loops to the extent that some West Bank locations are effectively isolated.
D. The plans show that the barrier will be constructed across the Dead Sea.
E. The barrier is very generally located along a path that parallels the 1949 Armistice Line.

Exploration 1.4: SHORT ESSAY

1. How does your perception of the site associated with the cultural icon placemark change after viewing it from this perspective? For example, is it bigger or smaller than you expected? Were you surprised by the surrounding landscape?

____
____
____

2. Think of another way that Google Earth could be utilized for law enforcement purposes. Provide a detailed description of your law enforcement tactic. Remember that Google Earth imagery is not real-time.

____
____
____

Name: ______________________________

Date: ______________________________

Chapter 2: Geography Concepts

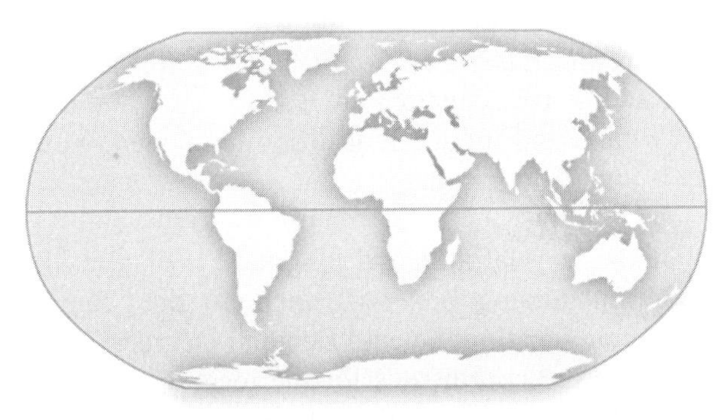

As you work through the explorations contained in this workbook, you will become well-versed in some of the basic concepts of the discipline of geography. You will hone your understandings of location, scale, and place via the assessment of remotely sensed data and information. The ability to effectively interpret this data and information is the key skill these exercises seek to develop.

Location is a concept that can be thought of in absolute or relative terms. For example, you are sitting *in front of* your computer is an example of relative location. On the other hand, absolute location refers to any system that provides us with a coordinate that represents a specific point on Earth's surface. The most commonly used method for determining location is the grid system of latitude and longitude. Latitude measures distance north and south of the Earth's equator, while longitude provides a measure of distance east and west of the Prime Meridian.

Another important concept to grasp is that of scale. When we use the term scale, we are referring to the relationship between units on a map or in this case, a digital globe, compared to units in the real world. Large-scale maps show a large amount of detail but not much area, whereas a small scale map shows a more extensive area with less detail. As you will experience in your work with Google Earth™, some types of analysis are better suited to viewing the Earth at larger scales, while others dictate a small scale approach.

Place is a final essential concept. Place refers to what makes one location unique from another. Place analysis is a subjective endeavor as the geographer tries to illuminate idiographic characteristics of a location. Oftentimes, these characteristics are based on social and cultural attributes such as economy, language, ethnicity, or religion.

The imagery that you view in Google Earth™ has all been captured remotely. Google Earth™ displays all kinds of different imagery that have been captured from remote sensors such as special cameras attached to satellites or airplanes. In this chapter you will be alerted to some potential difficulties of examining data that has been compiled from different sensors.

Finally, we will view a collection of locations that will help you develop some of the cognitive tools geographers use to evaluate and analyze the Earth from above. There are a number of "clues" present in any landscape that can help you understand and interpret the scene. These clues include things like the color, shape, texture, and size of objects in the field of view. Armed with this knowledge, you will be ready to explore world regions with Google Earth™.

Download EncounterWRG_ch02_GeographyConcepts.kmz from **www.mygeoscienceplace.com** *and open in Google Earth™.*

Exploration 2.1: LOCATION, SCALE, AND PLACE

Open the LOCATION, SCALE & PLACE folder in the *Geography Concepts* folder and double-click the *point of origin* placemark. What is significant about this location off the west coast of Africa? This is the origin for our globe's predominant location system. This is where the Earth's east/west baseline (the Prime Meridian/International Date Line) and the Earth's north/south baseline (the equator) intersect. Let's make it a little easier to see. Click "View" and then click "Grid." The latitude and longitude grid is illuminated. You can see that the point of origin placemark is located at 0° latitude and 0° longitude.

Now let's make sure that we are on the same page in terms of the way latitude and longitude is displayed. Click "Tools," and then "Options." The Google Earth™ Options dialog box will open. In the "Show Lat/Long" box, you will see four options for displaying your absolute location. Select "Decimal Degrees" and then click "OK." At the center bottom of your screen you will see the coordinates for the location of your cursor (the white hand). Move it around the globe. Examine what happens to the latitude and longitude numbers as you move to the southern and northern hemispheres and the eastern and western hemispheres. A positive latitude number represents locations north of the equator while a positive longitude represents locations east of the Prime Meridian. Move your cursor across the grid again to verify these statements.

Now manipulate the globe by rotating it and turning it to explore latitude and longitude at the poles and in the Pacific Ocean. Determine the maximum values for latitude and longitude and where those are located.

Exploration 2.1: MULTIPLE CHOICE

1. Which of the following coordinate pairs is associated with Lisbon, Portugal?

A. 38° North latitude, 7° West longitude
B. 38° South latitude, 7° West longitude
C. 38° North latitude, 7° East longitude
D. 38° South latitude, 7° East longitude
E. 7° North latitude, 38° West longitude

2. Based on your assessment of latitude and longitude, which of the following statements is incorrect?

A. The maximum value of longitude is 179.999°
B. The North and South Poles are located at 90° north and south latitude.
C. All locations south of the equator have negative values of latitude.
D. All locations located east of the Prime Meridian and west of the International Date Line (Antemeridian) will have east longitude.
E. The equator represents the 0° line of latitude.

3. Based on your assessment of latitude and longitude, which of the following statements is incorrect?

A. Lines of latitude are parallel with one another, thus insuring that a degree of latitude is equidistant anywhere on the globe.
B. All of Europe is classified as north latitude.
C. New Zealand's longitudinal values are higher (more eastern) than Australia's.
D. The Prime Meridian represents the 0° line of longitude.
E. Lines of longitude are parallel with one another, thus insuring that a degree of longitude is equidistant anywhere on the globe.

4. What physical feature is located at -15.72° latitude, 29.35° longitude?

A. mountain
B. island in the ocean
C. river channel
D. peninsula
E. volcano

Turn off the latitude longitude grid now to de-clutter the screen, but remember you can always turn it back on if you find it helpful or necessary to solve a problem. Double-click the *Mt. Kilimanjaro* placemark. You'll see that Mt. Kilimanjaro is located in the vicinity of -3° latitude, 37° longitude. As a famous mountain, it would be helpful to get a grasp on the elevation of the mountain. Turn on the *Terrain* layer in the *Primary Database*. Now when you move your cursor across the landscape you will see that the values for elevation displayed at the bottom of the screen will vary. Notice how the elevation decreases as you move away from the summit of Kilimanjaro. Zoom in and identify the highest point on Mt. Kilimanjaro. Remember that you will need to have the "Meters, Kilometers" option set for displaying elevation.

5. The summit of Mt. Kilimanjaro exceeds

A. 15,000 meters.
B. 7,300 meters.
C. 5,800 meters.
D. 2,900 meters.
E. 1,500 meters.

Now double-click the *Mt. Kilimanjaro 2* placemark. Notice that this perspective of the same location gives you a better visual appreciation of the relief in the area of Mt. Kilimanjaro. Sometimes it might be helpful to exaggerate the variations in elevation in order to see patterns on the landscape. Click "Tools," and then "Options." The Google Earth™ Options dialog box will open again. In the "Terrain Quality" box change the "Elevation Exaggeration" to "3" and then click "OK." You will need to back out to see all of Kilimanjaro. Beyond the mountain being more dramatically represented, notice how the stream channels on the flanks of Kilimanjaro are now more visible. Rotate the scene 360° by dragging the "N" in the compass. You will see other prominent mountains. Find the highest peak within approximately 100 kilometers.

6. The highest peak in the vicinity (<100 kilometers) of Mt. Kilimanjaro is located to the

A. north.
B. south.
C. east.
D. west.
E. northeast.

Return to the Google Earth™ Options dialog box and reset the "Elevation Exaggeration" to "1." Double-click the *Denver 1* placemark. You see the greater Denver metropolitan area. It's difficult to pick out many features of the metropolitan area at this scale. You should be able to identify the mountains to the west of town, some of the stream networks, agricultural regions, and the urban area. This is the smallest scale view of Denver that we will utilize. Now double-click the *Denver 2* placemark. This view is a larger-scale view than the one associated with *Denver 1*. By zooming-in we can see much more detail, but we've lost some of the big picture. For example, the high mountains are no longer visible. But now we can see highways and the major roads and features like reservoirs and green space. Move to the *Denver 3* placemark. Much has been gained by zooming in. This is a larger-scale view than *Denver 2*, but will be smaller scale than *Denver 4*. Now we see features such as sports stadiums, highway interchanges, and the downtown area. However, we have no idea how big the metropolitan area is and we have no way of knowing that there are mountains immediately to the west.

7. What *cannot* be verified by the Denver 3 view?

A. The major transportation corridor visible generally runs north/south.
B. The downtown area street grid is set on a 45° angle.
C. There are large football and baseball stadiums in the view.
D. A river roughly parallels the primary transportation corridor.
E. The major transportation corridor does not have bridges in the view.

Double-click *Denver 4* and you will see a level of detail that enables you to gain a very in-depth understanding of the features on the ground. Here we see an amusement park and its parking lot along with a river. Large-scale images like this facilitate analysis and interpretation. For example, we could use this image to determine the capacity of the parking lot or the spatial distribution of crosswalks or the amount of the park that was shaded by trees.

This rich detail begins to give us a sense of this place. Clearly, this is a society with a relatively high level of affluence as evidenced by the abundance of recreational facilities. Google Earth™ has a number of layers available that can help us develop a sense of place. One of the best is the 360° panoramic images. Fly into the *Denver street-view*. Manipulate the view to look up, down, and all-around.

8. Based solely on the Denver street-view, what quick judgment is not supported?

A. Property in this vicinity is very valuable as evidenced by the vertical development.
B. Mass transit is available.
C. There is no evidence that individual vehicles are utilized to any significant degree.
D. There is likely a Christian population in the immediate area.
E. The climate here is temperate.

For a comparison, fly into the *Shibam street-view*. Clearly this urban snapshot of life in Yemen is a vastly different scene than what you explored in the Denver street-view.

9. Based solely on the Denver and Shibam street-views, what quick judgment is not supported?

A. The climate in Shibam is likely arid.
B. The Shibam image is an area visited by tourists.
C. Infrastructure such as roads and the electrical grid is less developed in Shibam.
D. Some women in Shibam wear very modest attire.
E. There is an emphasis on modern architecture in Shibam at the expense of folk architecture.

Exploration 2.1: SHORT ESSAY

1. Use Google Earth™ to determine the latitude and longitude of your university and your home. List the coordinates for both and explain why the numbers are lower or higher in comparison with one another.

____

____

____

2. Return to the Denver 4 placemark and study the image. Provide three additional questions that could be explored with this imagery (e.g., what's the capacity of the parking lot).

____

____

____

Exploration 2.2: REMOTELY SENSED DATA

As discussed in the introductory text of this chapter, Google Earth™ makes extensive use of remotely sensed imagery. For the purpose of this text, it's not critical that you understand the intricacies of the sensors involved, but rather that you are aware of the nuances of remotely sensed imagery that is utilized by Google Earth™. This exploration will help you understand some of the issues you are likely to encounter when working with Google Earth™. Open the REMOTELY SENSED DATA folder and double-click the *resolution 1* placemark. When we use the term resolution we are referring to the spatial resolution or the measurement of the minimum distance between two objects that will allow them to be differentiated from one another in an image. A higher resolution image would have a lower number associated with it. For example, a 30-meter spatial resolution image would be a higher resolution image than a 100-meter spatial resolution image. The imagery to the right of the *resolution 1* placemark has a higher spatial resolution than the imagery to the left. Notice how individual features, such as trees, are easier to discern on the right. Zoom in and see how you can see the intricacies of the shoreline in the body of water on the right compared to the image on the left where features are much more generalized. Examine the *res a* through *res e* placemarks, zooming in as necessary.

Exploration 2.2: MULTIPLE CHOICE

1. Which of the following placemarks is located in the area with the *lowest* spatial resolution imagery?

A. res a
B. res b
C. res c
D. res d
E. res e

You just had a taste of the way that different image sets overlapping can create different impressions of a landscape. These different sets can be caused by the fact that images have been captured by different sensors, have been taken at different times of the year, or were captured in different years. For example, double-click the *imagery sets* placemark. The varied imagery makes this region look very differently. Without zooming in, how many imagery sets can you identify in this view?

2. How many imagery sets does the view associated with the imagery sets placemark contain?

A. 1
B. 4
C. 8
D. 12
E. >15

Go to the *seasonal differences* placemark. You will see distinct differences between the landscape on the right and left sides of the screen. Whenever you are considering the climate, agriculture, or vegetation of a scene, be sure you keep in mind the seasonal considerations. You can do this by using your common sense. For example, do the deciduous trees have leaves on them or not? Additionally, you can usually find out the date the image was captured by moving your cursor over the landscape. The imagery date(s) will appear in the bottom left-hand corner of the screen. Additionally, be wary of the fact that Google will sometimes add color to imagery to help smooth the edges of image sets. This is quite apparent in one of the images associated with this placemark. Be careful not to let this bias your impression of a landscape.

3. Which of the following statements regarding the seasonal differences placemark is not accurate?

A. The imagery on the left was captured in the spring.
B. The imagery on the right has had color added.
C. The imagery on the right was collected in January.
D. The imagery on the left was captured in 2004.
E. There is evidence of agricultural activity on both sides of the scene.

The variations in imagery captured in different seasons can also impact your ability to interpret built environments. Significant urban forests can obstruct the views of communities from above. Go to the *leaf-on* placemark and explore the image of a city captured in late spring. Then go to the *leaf-off* placemark. The built environment is much more visible in the leaf-off placemark. However, the location does not look as aesthetically appealing. Remember this factor as you interpret landscapes throughout this text.

Now double-click the *image variation by date* placemark. This is a case where you have two images of different vintages abutting one another. Evaluate the scene and think about what kinds of problems could arise because of this.

4. Which of the following evaluations of Heathrow Airport could be completed with the available imagery?

A. Total number of aircraft on the ground on March 5, 2006.
B. Total area of Heathrow structures on March 5, 2006.
C. Total area of Heathrow structures on September 28, 2008.
D. Length of Heathrow's southernmost runway on September 28, 2008.
E. Total number of aircraft on the ground on September 28, 2008.

Double-click the placemark, *imagery through time*. You will zoom to an area near Birmingham, Alabama. It is important for you to realize that you are usually not resigned to viewing an area of the Earth at only one snapshot in time. The historical imagery function of Google Earth™ can be very helpful in highlighting change that takes place over time. In the Google Earth™ toolbar, click the historical imagery button (it has a picture of a clock). A time slider will appear in the upper left-hand corner of the image. Use this slider to see the landscape at various points in time. As you work through this text, you will not always be reminded to use this tool, but try not to forget about it as it can make your life much easier.

5. Which of the following statements is not supported by the historical Google Earth™ imagery?

A. High resolution imagery for the area is available in Google Earth™ dating back to 1978.
B. Leaf-on and leaf-off imagery is available in the historical series.
C. Buildings are constructed at the site A placemark sometime between 1998 and 2002.
D. Site B has forest cleared between 2005 and 2007.
E. No roads are constructed at site C between 2005 and 2006.

Turn off the historical imagery by clicking the clock icon on the toolbar.

Exploration 2.2: SHORT ESSAY

1. What types of problems could arise when interpreting a scene that contains imagery from different years? Specifically, provide some examples of changes that can occur in both the physical and cultural landscape over time.

____
____
____

2. What are the most noticeable changes you can identify in the vicinity of the imagery through time placemark? How would you describe development in this region? Is it happening rapidly, moderately, or not at all? Support your answer with examples from the historical imagery.

____
____
____

Exploration 2.3: INTERPRETATION

Throughout the text you will be asked to problem solve by identifying unique features or providing some type of interpretive analysis of cultural and physical landscapes. There are several elements of interpretation that are usually present to some degree in any imagery. These elements are size, shape, texture, pattern, association, shadow, and site/situation.

Tone and/or color is one of the most distinguishing characteristics of a given feature in Google Earth™. Open the *tone/color 1* placemark and you will see a large football stadium. The lush green of the pitch stands out. Another field might not be so well-maintained and would not be the deep, even green we see at Camp Nou, home to FC Barcelona. We also see the distinctive red roofs of the region, the gray of asphalt, and the numerous colors of automobiles. Now open *tone/color 2* and view the coastal waters. The varied shades of blues and greens can indicate the presence of subsurface vegetation or coral structures. Here, the darker colors are related to deeper channels. *Tone/color 3* presents a homogenous scene of color.

Exploration 2.3: MULTIPLE CHOICE

1. The tone/color 3 placemark is centered in an area with relatively uniform color. This is a landscape dominated by

   A. cotton fields.
   B. forest.
   C. sand.
   D. snow and ice.
   E. water.

Size can help your interpretation in relative and absolute ways. Open the *size 1* placemark and you will see Boone Pickens Stadium at Oklahoma State University. We can obtain an absolute size from this image because we know that an American football field is 120 yards (about 110 meters) long if you include the end zones. You can apply this knowledge to measure the structure next to the stadium, Gallagher-Iba Arena. We also see opportunities for relative size evaluation. Automobiles vary in size somewhat, but they can provide a comparative reference point.

Zoom to the *size 2* placemark to see another really good example of an absolute size marker. The vast majority of railroad tracks in the US are a standardized gauge with the rails spread 4' 8.5" (1.48 meters) apart from one another. Open the *size 3* placemark.

2. If you wanted to determine the absolute size of the reservoir in the picture, what feature visible in the scene would be the best choice to use as your size reference point?

A. baseball field
B. parking lot
C. softball field
D. tree crown size
E. tennis court

Double-click the *shape 1* placemark. You will fly to an industrial agriculture landscape. These are center-pivot irrigation schemes on the Great Plains. They utilize groundwater pumped to the surface and distributed by a sprinkler that rotates around a fixed pivot in the center of the field. Moving on to the *shape 2* placemark, you will likely recognize this feature as it is named after its shape. The Pentagon in Washington D.C. is one of the world's largest structures. The *shape 3* placemark has some distinctive linear shapes. Explore this scene and identify the features.

3. The linear features associated with the shape 3 placemark are

A. airport runways.
B. interstate highways.
C. streets in a housing development that has yet to be built.
D. skyscrapers.
E. railroad tracks.

The texture of a surface can be quite fine or very coarse. This can give you ideas about the types of vegetation on the ground or about the degree of homogeneity of a landscape. Open *texture 1* and view the dense forest. This is a relatively uniform landscape. *Texture 2*, on the other hand, has a significant degree of variation. You are viewing an ecotone as the landscape transitions from steppe into woodlands and forest from east to west. This is a result of changing elevation and precipitation. Evaluate the *texture 3* landscape.

4. Based upon your evaluation of the texture 3 landscape, identify the *most* accurate statement.

A. This is a landscape with no variation in texture.
B. This landscape has significant variation in texture as a result of some agricultural activity.
C. Texture varies in a regular and geometric pattern.
D. The western side of this scene has much more variation in texture than the eastern side.
E. This landscape has high texture variation because of its combination of vegetation and the built environment.

More often than not, regular patterns on the landscape suggest human involvement. For example, the *pattern 1* placemark represents natural vegetation that is being intensively managed by humans. In this case, we are looking at one of the world's largest pecan orchards. The *pattern 2* placemark highlights the concentric rings associated with surface mining. As this mining operation continues, this pattern grows in depth and width (at the expense of the existing community). Now zoom to the *pattern 3* placemark to view a pattern that is common across the western US. This is a planned suburb where the houses are all very similar in size and appearance. In the scene there are only two buildings that deviate significantly from the pattern. Zoom in and determine what these structures are.

5. The structures that deviate from the pattern in the pattern 3 scene are

A. homes for the very wealthy.
B. retail establishments.
C. schools.
D. factories.
E. hospitals.

Association refers to the ability to identify or confirm the existence of a feature based on its relationship to other features. Double-click the *association 1* placemark and you will find a collection of long yellow vehicles in a parking lot. These are school buses. Why would you have that many school buses in one location? Perhaps this is the bus yard for a large metropolitan school district? When we zoom out we see that it is not a large city, but there is a large industrial building adjacent to the buses. A little research reveals that this is Fort Valley, Georgia, home of the Bluebird Bus Company factory. Zooming to the *association 2* placemark reveals strange groupings of what appear to be parts of airplanes. Let's zoom out and see if there is anything that would help confirm this. In fact, there are numerous intact airplanes and also an airport nearby that explains how this material has arrived. By the way, this is one of the US military's airplane bone yards. They are usually found in desert areas, because the planes do not rust as quickly in the dry climate. Zoom to *association 3* and use the associate features to help you determine what we are viewing.

6. The association 3 placemark is

A. a railroad bridge.
B. a dam.
C. a combo rail and highway bridge.
D. an aqueduct.
E. a highway bridge.

Shadows can be helpful in providing an idea of a feature's height and shape as it can sometimes be difficult to ascertain this from a vertical perspective above the object. For example, the *shadow 1* placemark illustrates Houston, Texas. Downtown is most distinguishable from its buildings' shadows. Excluding a few tall buildings due west of downtown there is little vertical development of real estate outside the central business district. The *shadow 2* placemark illustrates a distinctive shadow associated with a cooling tower at one of France's largest nuclear power plants. Something is odd about this image though. Notice that cooling tower is grainier than the rest of the image. Also, where is the shadow for the second cooling tower? This imagery has been modified/distorted for security reasons. Now go to the *shadow 3* placemark in Washington D.C.

7. Provide your best estimate as to what time of day this image was captured.

A. 8 AM
B. noon
C. 2 PM
D. 5 PM
E. 8 PM

Site/situation refers to the geographic context of the feature, features, or landscape that you are interpreting. For example, go to the *site/situation 1* placemark and you will see an airport with a number of planes and helicopters positioned on the tarmac. You also see a large number of modular structures around the specific location (the site). Whereas site refers to the local circumstances, situation refers to the regional context. The situation of this image explains the aircraft and structures. If we back out and turn on the *Borders and Labels* layer in the *Primary Database* you will see that we are in Afghanistan. This is a military base supporting the war efforts there. *Site/situation 2* takes us to Iceland and the volcano that shut down air traffic over much of Europe in April of 2010. At the time this text was written, Google did not have updated imagery illustrating the eruption. As you look at this placemark, it is possible that has changed. The moral of the story is: Google Earth™ imagery is not live. Sometimes you will find imagery that is very recent, while other times the imagery may be more than five years old. *Site/situation 3* also underscores the necessity to be aware of contemporary contexts. We have zoomed to the grounds of a Presidential Palace. If this is a formal location, why are thousands of people camped out all around the site?

8. Site/situation 3 illustrates

A. the site of a recent earthquake.
B. Earth Day.
C. a massive demonstration against the government.
D. the site of a recent tsunami.
E. a street festival.

Exploration 2.3: SHORT ESSAY

1. Select one of the interpretation placemarks and describe the evidence of all of the interpretive elements visible in that scene.

____
____
____

2. What two elements of interpretation do you think are most helpful to you when trying to understand Google Earth™ imagery? Explain your answers and provide examples.

____
____
____

Name: ____________________

Date: ____________________

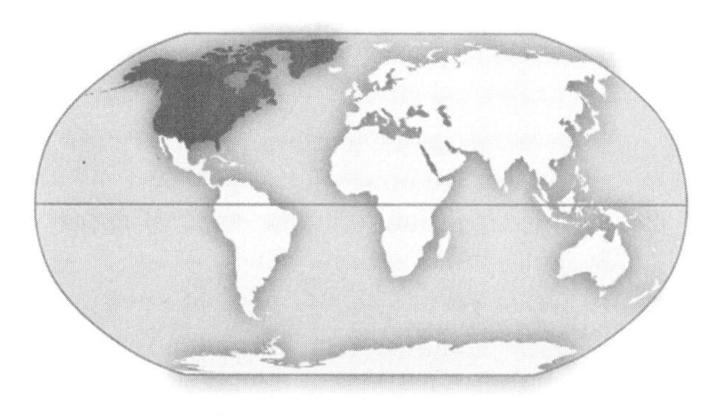

Chapter 3: North America

North America, the northern continent of the Americas is bounded by the Arctic Ocean on the north, the North Atlantic Ocean on the east, the Pacific Ocean on the West, and the Caribbean Sea on the southeast. Physically, it is reasonable to include Mexico and Central America. However, cultural differences dictate that those areas are considered part of Latin America for the sake of these explorations. The most populous country in the region is the United States with approximately 309 million in 2010. Canada's 2010 population is 34 million. The most populous metropolitan area is centered on New York City and has a population near 20 million.

North America is home to varied types of energy development such as oil and gas, coal, and hydropower. Wind energy is becoming an increasingly popular "green" source of electricity. High oil prices have not only led to the development of environmentally friendly sources of energy. For example, recent price spikes have supported the increased development of tar sands in Canada. With ongoing economic growth and development, the demand for energy will continue to grow. Producing energy domestically can help meet these needs and also enhance national security by lessening dependence on foreign sources.

The ever-growing demand for energy resources can be better understood by surveying North American cities. Cities in this region are often spatially extensive, sprawling entities that present numerous challenges to efficient transportation and environmental well-being.

One ubiquitous component of urban sprawl is the shopping mall. Some of the world's largest shopping malls are located in the region. However, shopping mall forms and functions have evolved with consumers' changing preferences. This evolution can be seen from the birds-eye perspective of Google Earth™.

Another conspicuous element of the landscape that can be viewed from this perspective is the footprint of the US military. The varied purposes of the bases are reflected in both the built and natural environment of military installations. The reach of the US military extends beyond these sites, however, as service men and women are stationed around the world.

This region is home to a range of agricultural production methods that produce a wide array of products for consumers in the region and around the world. The scale of these operations, ranging from the transportation and storage components of the chain of supply to the fields themselves are visible elements of this region's landscapes.

Download EncounterWRG_ch03_NorthAmerica.kmz from ***www.mygeoscienceplace.com*** *and open in Google Earth™.*

Exploration 3.1: NORTH AMERICA ENVIRONMENT

Let's tour some of the sites of energy development around the North America region. The visible "footprints" of energy production often provide conspicuous signatures.

Exploration 3.1: MULTIPLE CHOICE

Go to placemark *A*. Zoom in and out as necessary and do some research to find out what type of energy production is common in the area.

1. What are the "dots" on the landscape surrounding placemark *A*?

A. geothermal bores
B. cold fusion reactors
C. wind turbines
D. natural gas wells
E. coal mine shafts

Placemark *B* highlights a hydroelectric facility. In the middle part of the 20th century, dams were often viewed as a more environmentally friendly way to produce energy. Today, this viewpoint is not as widely held. Study the image and determine what feature(s) most clearly represent an attempt to mitigate an adverse environmental impact associated with the dam. It will be helpful to view some of the *Panoramio Photos* (check the box next to *Panoramio Photos* in the *Primary Database* in *Layers*) around this location. After you have viewed the photographs it's always a good idea to turn-off the *Panoramio Photos* because in certain situations they will really clutter your view.

2. What feature(s) around placemark *B* did you identify as an attempt to mitigate an adverse environmental impact?

A. The river heaters designed to heat the cool pool of water backed up behind the dam.
B. The diversion of the river into slow, medium, and fast channels.
C. The plastic vegetation and trees to simulate a less-disturbed environment.
D. The fish ladders to help migrating fish circumvent the dam.
E. The dirt parking lot to minimize run-off.

Placemark *C* illustrates the utilization of wind for energy production. Let's think about why these features are located here. You should be able to get an idea by the dark areas on the map. These are stream channels. Their density and direction suggest that there is a local change in relief (the difference between lower and higher elevations in a given area). You can accentuate this difference in a couple of ways. First go to "Tools" and then "Options" and change the vertical exaggeration factor to "3." Be sure the *Terrain* layer in the *Primary Database* is on. When you explore the area around the turbines, you should be able to see the local differences in elevation more clearly. Remember to tilt your view to better see changes in elevation. A second way to observe this is to click the "show sunlight across the landscape" button. It's the button with a picture of a small sun partially behind a hill. Use the time slider to simulate the movement of the sun across the sky. The shadows help you determine the areas of greater elevation differences. Be sure to reset your elevation exaggeration to "1" when you have completed your exploration of the site.

3. The wind turbines around placemark *C* are located

A. adjacent to standing water where they can capture evaporative currents.
B. in a valley where they are most likely to take advantage of rapidly rising or sinking air.
C. on a ridge where they are most likely to take advantage of rapidly rising or sinking air.
D. next to an interstate highway where they can utilize the air generated by large vehicles.
E. where there are no trees to interfere with air flow.

Placemark *D* is an area of coal mining in West Virginia. At this location the technique of "mountaintop removal" is employed. In the imagery you can see locations that are in stages from initial clearing to digging to reclamation. Think about what order you would place the numbered placemarks to reflect the sequence: clearing, digging, reclamation. If you are interested in learning more about this process, turn on the *Appalachian Mountaintop Removal* layer in the *Global Awareness* folder of Google Earth™.

4. Order the numbered placemarks around placemark *D* to reflect the sequence of clearing, digging, reclamation.

A. 1,2,3
B. 2,1,3
C. 3,2,1
D. 1,3,2
E. 2,3,1

Placemark E is the Athabasca Oil Sands in the Canadian province of Alberta. These are deposits of very heavy crude oil that are surface mined. The process is expensive, but high oil prices have contributed to an expansion of these operations since 2004. Examine the 1974 and 2004 image pair, by turning them on and off, respectively. Now turn off both images and view the imagery available from Google Earth™.

5. The imagery from Google Earth™ appears to have been captured

A. prior to 1974.
B. sometime between 1974 and 2004, but closer to 1974.
C. sometime between 1974 and 2004, but closer to 2004.
D. following 2004.
E. from an entirely different location.

Exploration 3.1: SHORT ESSAY

1. Explain the factors that led to your conclusions regarding the approximate date the Google Earth™ imagery was captured. What features on the landscape changed through time? If you wanted to answer with complete certainty, what kinds of measurements could you complete? What are some of the factors that complicate the comparison of the different images?

____
____
____

2. Compare and contrast the varied imprints on the landscape of the five forms of energy extraction/production you have viewed. Which form seems to have the least environmental impacts? What are some adverse environmental impacts that could be associated with this form that are not readily apparent from the imagery?

____

____

____

Exploration 3.2: NORTH AMERICA POPULATION

Try to get in the habit of turning off the previous explorations layers and placemarks by un-checking the box next to the respective folder when you have completed the exploration. It will help keep the clutter from your view and will turn-off layers that can potentially obstruct your view for the next exploration. In this case, verify that the box next to 3.1 ENVIRONMENT is empty (the folder is turned off), before beginning this exploration.

Oklahoma City, Oklahoma is the largest (by area) city in the United States (excluding consolidated city-counties). Zoom in and explore the OKC metropolitan area by double-clicking the *Oklahoma City, OK* placemark. Then use the ruler tool to measure the approximate area of the city with its adjacent suburbs. The combined statistical area of Oklahoma City has a population of approximately 1 million persons. One can calculate population density by dividing population by area. Calculate an approximate figure for population density of Oklahoma City.

Exploration 3.2: MULTIPLE CHOICE

1. What figure is closest to the approximate area of the Oklahoma City, OK metropolitan area?

A. 2 km^2
B. 20 km^2
C. 200 km^2
D. 2,000 km^2
E. 20,000 km^2

Type "Tokyo Prefecture" in the fly-to box of Google Earth™. A prefecture is Japan's first level of sub-national entities. This is similar to what would be referred to as a state in the United States. Japan has 47 prefectures. Now determine the approximate area and population for Tokyo Prefecture, also known as Tokyo Metropolis, and calculate the approximate population density. You can do this by making general measurements in Google Earth™ or locating more explicit data from outside sources. Compare your calculations with what you know of Oklahoma City.

2. Population density in the Tokyo Metropolis is approximately

A. half that of OKC.
B. the same as OKC.
C. twice that of OKC.
D. five times that of OKC.
E. 20 times that of OKC.

Examine the 2nd and 3rd largest cities in the US by area (*Houston, TX* and *Phoenix, AZ* placemarks). Study the physical geography in the immediate vicinity of each city (Think about physical constraints to continued spatial expansion.)

3. If OKC, Houston, and Phoenix had similar gains in population, what city's population density would be the most likely to increase? Answer this with the physical geography of the three metropolitan areas in mind.

A. OKC
B. Houston
C. Phoenix
D. OKC and Houston
E. OKC and Phoenix

Let's take a look at the staggering growth that has occurred in and around Las Vegas in the last 30 years. Double-click the *Las Vegas LANDSAT* folder, and then expand the folder. Toggle the 1973, 2000 and 2006 LANDSAT images on and off to view the changes. Beyond the impressive expansion of the city, you will notice a significant increase in bright green patches. This is a surprising find in the desert. Determine what these features are.

4. What are the bright green patches visible in the Las Vegas LANDSAT imagery (particularly in the west and south sides of the city) that have increased notably from 1973 to the present?

A. golf courses
B. large public parks
C. football fields
D. lawns
E. algae-covered ponds

Turn off the *Las Vegas LANDSAT* folder. Now compare the *Las Vegas neighborhood* placemark and the neighborhood associated with the *Sprawl Example* placemark. Which neighborhood has a higher population density and why?

5. Comparing the Las Vegas neighborhood and the Sprawl Example, which neighborhood has a higher population density and why?

A. Las Vegas because most structures in the view are multi-story.
B. Sprawl because the houses are larger and likely have more people living in them.
C. Las Vegas because the lot size for each home is much smaller.
D. Las Vegas because the houses are larger and likely have more people living in them.
E. Sprawl because the lot size for each home is much smaller.

Exploration 3.2: SHORT ESSAY

1. Compare and contrast the urban environments of OKC and Tokyo. Identify at least three visual cues that would verify that the population density of Tokyo is higher than that of Oklahoma City.

____
____
____

2. Look at the placemark titled Sprawl Example and identify features of this scene that make it a good example of urban sprawl. Use your textbook for help with appropriate terminology. Also consider the wider context, such as where in the US this example is located and its proximity to higher population density areas.

____

____

____

Exploration 3.3: NORTH AMERICA CULTURE

Examine the first four shopping malls that are placemarked in the *Culture* folder. Note similarities and differences between these malls.

Exploration 3.3: MULTIPLE CHOICE

1. Which of the following is not a common component of the first four malls placemarked in the Culture folder?

   A. Large surface parking lots and/or parking garages
   B. Mass transit hubs in the form of subway/rail and/or bus stations on the premises
   C. Additional commercial development nearby
   D. Larger "anchor store" structures located at opposite ends of the mall structure
   E. Nearby (<1km) access to an arterial highway

Open the link associated with the Mall of America. Examine this list of the largest North American malls.

2. Based on opening dates provided by the link attached to the Mall of America placemark determine what time period can best be described as the era of the mega-shopping mall.

   A. 1930-1950
   B. 1940-1960
   C. 1950-1970
   D. 1960-1980
   E. 1990-2010

If the era of the mega-mall has drawn to a close in the United States, does this suggest that malls are no longer being built? In fact, enormous shopping malls are popping up in Asia because of inexpensive land and cheap labor. Now most of the world's largest malls can be found in places like China, Malaysia, and the Philippines.

Click the placemark for the *Golden Resources Mall.* This mall opened in 2004 and is one and one half times the size of the Mall of America.

3. When you can compare the Golden Resources Mall and its immediate setting with its counterparts in North America, what is the most noticeable difference?

   A. Golden Resources has noticeably fewer surface parking spaces.
   B. Golden Resources does not have nearby (<1km) access to an arterial highway.
   C. Golden Resources does not have nearby (<1km) residential areas.
   D. Golden Resources is only one story tall.
   E. Golden Resources does not employ sky lights to allow natural light into the mall.

Returning to North America, a major trend in the last decade has been the development of lifestyle centers. Lifestyle centers are usually located in affluent suburbs that offer a number of upscale shopping amenities. Examine the *The Promenade Shops at Centerra* and note differences in comparison to the large malls you have already examined.

4. Select the most logical statement based on your observations of the malls and lifestyle center.

A. Lifestyle centers can offer quicker access to stores for customers and lower cooling and heating costs for operators.
B. Lifestyle centers are desirable to consumers because shoppers can move from store to store in a climate-controlled environment.
C. Lifestyle centers are designed to cater to the North American mentality that bigger is better.
D. Lifestyle centers require larger development sites than traditional malls.
E. Lifestyle centers would most likely be built in densely built-up inner-city areas.

The Mall of America boasts 390,000m^2 of *floor* space. Surprisingly, one of these: *Building A*, *Building B*, *Building C*, *Building D*, or *Building E* has more than 600,000m^2 of floor space. Use the ruler tool and your critical thinking skills to determine which building is the one with the greatest floor space. It may also be helpful to turn on the 3D Buildings layer.

5. Which building is largest, in terms of floor space?

A. Building A
B. Building B
C. Building C
D. Building D
E. Building E

Exploration 3.3: SHORT ESSAY

1. What factors can explain the trend of new shopping centers being more likely to be a lifestyle center rather than a large, enclosed mall? Consider general cultural factors, planning strategies, level of affluence, and mass transit in your response.

____
____
____

2. What are the five buildings associated with the placemarks *Building A*, *B*, *C*, *D*, and *E* and how are they utilized? Attempt to rank them by floor size and by total volume. Provide a few sentences of explanation for your rankings.

____
____
____

Exploration 3.4: NORTH AMERICA GEOPOLITICS

Military installations in the United States can provide prominent landscapes. The primary training and/or deployment purposes may be evident by examining these installations in Google Earth™. For example, examine the *Norfolk Naval Station* placemark. This is the largest naval station in the United States. You can see aircraft carriers, submarines, destroyers, and assorted support vessels.

Exploration 3.4: MULTIPLE CHOICE

1. Can you identify how many aircraft carriers were in port at Norfolk when this image was captured?

A. 0
B. 3
C. 5
D. 7
E. 9

Now take a look at two of the largest US Army bases, *Ft. Hood* and *Ft. Benning*. Specialized training takes place at both these locations. Explore the bases and their surrounding environments. Now examine the area in and around *Edwards Air Force Base*. Again turn on the *Panoramio Photos* in the *Primary Database* and view pictures of the site.

2. Select the statement that most likely explains key training missions at Ft. Hood and Ft. Benning.

A. tanks at Hood and paratroopers at Benning
B. amphibious assault at Hood and tanks at Benning
C. tanks at Benning and desert warfare at Hood
D. amphibious assault at Benning and paratroopers at Hood
E. paratroopers at Hood and desert warfare at Benning

3. Regarding Edwards Air Force Base, which of the following statements is not accurate?

A. The base is not located in an urban area.
B. The space shuttle occasionally uses this site as a landing strip.
C. The Sierra Nevada Mountains are less than 10 kilometers to the west and thus provide a natural barrier to high winds that could interrupt flight tests.
D. The landscape suggests that there is not much rain, and therefore an abundance of clear weather.
E. The base is located adjacent to a dry lake bed that could be utilized as an extension to the runway.

The United States also has a significant military presence overseas. Some of these service men and women are stationed on aircraft carriers like we viewed at Norfolk, while others work at large overseas bases, like Ramstein in Germany. Click the *US Active Duty Military* layer and explore the globe. The countries with the 15-highest counts of active US military in October 2009 are highlighted. These numbers are in a state of continuous flux, particularly in areas where active combat operations are taking place. You can zoom in to each extruded polygon to see a label with the troop count. Keep in mind that these counts do not include US Naval fleets around the world. In October 2009, there were more than 20,000 active personnel afloat outside of the United States.

4. Utilizing the US Active Duty Military layer, determine what country had the fourth-highest count of US military personnel in October 2009.

   A. Iraq
   B. Japan
   C. Germany
   D. Korea
   E. Italy

5. US military personnel are exposed to a wide range of cultural variables in their diverse work places. In which of the top 15 countries would someone be most likely to have machboos at dinner?

   A. Japan
   B. Germany
   C. Turkey
   D. Bahrain
   E. Portugal

Exploration 3.4: SHORT ESSAY

1. Did any of the top 15 troop locations in the US Active Duty Military layer surprise you? Identify the top five and briefly describe why the US has such a large presence at each of these locations.

   ____
   ____
   ____

2. Return to the four bases in the US that were placemarked and evaluate the base and its surroundings. What impacts on the bases' immediate and surrounding physical and cultural landscapes are evident? Consider facets like environmental impacts and residential and commercial development in your response.

   ____
   ____
   ____

When you have completed Exploration 4, turn off the GEOPOLITICS layer.

Exploration 3.5: NORTH AMERICA ECONOMY AND DEVELOPMENT

Expand the *ECONOMY AND DEVELOPMENT* folder and examine the *West Texas* placemark. The unique pattern you see is provided by a type of irrigation known as center pivot irrigation. In this part of Texas, they produce crops such as cotton and wheat. The mean annual precipitation in this part of North America is insufficient to grow many crops. Zoom in and study the landscape to determine where the water is most likely being obtained.

Exploration 3.5: MULTIPLE CHOICE

1. Where is the water being obtained to water the fields you see in West Texas?

A. It is dropped from airplanes.
B. It is trucked-in daily.
C. The lack of surface water suggests it is likely groundwater.
D. It is piped from the numerous dammed reservoirs in the area.
E. It is diverted from streams in the area.

A great deal of the crops that are grown across the broad swath of North America's Great Plains and prairies ends up being shipped by rail and/or barge. Because of the size of the containers and their ability to be coupled, these are more efficient mechanisms to move these products than trucks. An important destination for the transport of a particular agricultural commodity in the central US can be seen near the *Wichita, Kansas* placemark. The yellow arrows are pointing at a feature that stores a particular type of agricultural commodity. Determine what that commodity would most likely be. You'll likely need to research this beyond what you can learn from Google Earth™.

2. What agricultural commodity is most commonly stored in the large structures highlighted with the yellow arrows at the *Wichita, Kansas* placemark?

A. corn
B. pork bellies
C. cattle
D. wheat
E. milk

Our next stop takes us to *St. Louis*. Here you can see a number of barges moving along the Mississippi River. Navigable rivers are the most efficient way to move very large amounts of grain from the agricultural heartland of North America. Wichita, Kansas is located on the Arkansas River. However, when we explored Wichita we could find a surplus of railroad activity, but no barges. When you click the *Arkansas River locks* placemark, you see proof that the river is navigable to barge traffic at some points by the evidence of locks to move river traffic from higher to lower flood plains or vice versa. The next placemark (*Arkansas River barges*) gives you tangible proof that barges are found on the Arkansas River. Then why in the world aren't the agricultural commodities from Wichita being moved by barge? That's correct; the river is not navigable in Wichita. Locate the farthest western navigable point for barge traffic on the Arkansas River or its immediate tributaries (to the nearest degree of longitude). Use evidence of barge-sized locks or better yet, barges themselves.

3. The farthest western navigable point (to the nearest degree of longitude) for barge traffic on the Arkansas River or its immediate tributaries is

A. 92° W.
B. 94° W.
C. 96° W.
D. 98° W.
E. 100° W.

Check out the *Feedlot A* placemark. This is a different type of industrial agriculture operation. This is a concentrated animal feeding operation with up to 150,000 cattle. These locations finish fattening up cattle before they are slaughtered. Can you guess what the dark areas inside the individual pens are? Look at another feedlot (*B*) in the area. Determine what competitive advantage one site has over the other.

4. What feedlot (*A* or *B*) has a competitive advantage over the other and why?

A. Feedlot A has onsite railroad access.
B. Onsite slaughtering and meat-packing at Feedlot B.
C. Feedlot B close to large area of surface water.
D. Onsite slaughtering and meat-packing at Feedlot A.
E. Feedlot A close to large area of surface water.

Turn off all of the layers in the *Primary Database*, and then click the "what's going on here" placemark. There are clearly some significant differences between the landscapes that are above and below the "line" running through the middle of the picture. With a little exploration you would be able to determine that this is an international border. In this case, we are looking at the US/Mexico border. Zoom in to the urban area at the center of the image and follow the border until you locate the major checkpoint between the countries. Is traffic backed-up coming into the US (leaving Mexico) or coming into Mexico (leaving US)?

5. Is traffic backed-up coming into the US (leaving Mexico) or coming into Mexico (leaving US) at the urban area at the center of the what's going on here placemark?

A. coming into the US
B. coming into Mexico
C. approximately the same level of traffic entering both countries
D. no traffic back-up on either side of the border
E. this is no border passage at this site

Exploration 3.5: SHORT ANSWER

1. Describe at least three differences between similar features on the respective sides of the US and Mexico border that can be noted at the original scale of the what's going on here placemark or at a larger scale (more zoomed-in). For each of the differences you note, provide a possible explanation for the difference.

____

____

____

2. You have been exposed to just a few examples of industrial agriculture in North America. Use Google Earth™ to locate another example. You might consider logging in the Pacific Northwest, aquaculture along the coasts, or fruit orchards in California or Florida. These are but a few of the other forms of industrial agriculture that provide unique cultural landscapes for us to examine in Google Earth™. Provide the coordinates of the location and a description of what you see. Be sure to provide the details that helped you determine what it is you have identified. Feel free to supplement your work with outside resources.

____

____

YOU MAP IT! Urban Sprawl and Centralization

Identify areas of urban sprawl and anti-sprawl in and around your city of residence or a city with which you have a high degree of familiarity. Utilize any of the tools of Google Earth™ to create a map of relevant features for the zones in the community of your choice. Examine the *You Map It!* example for Oklahoma City. Provide some descriptive text that explains the highlighted areas/features. Here's an example of text that corresponds with the example placemarks in the YOU MAP IT! – Urban Sprawl folder.

> Oklahoma City is a metropolitan area that has continued to sprawl in nearly every direction from the 1960s through the present. You can select virtually any peripheral location and examine the changes that have occurred through time. If you click the *Sprawl Example* for OKC, then use the time slider (the small clock in the Google Earth™ toolbar), big changes are evident. For each of the locations discussed, click the placemark, and then manipulate the time slider to see the changes take place.
>
> You can see that roads have been widened and upgraded (*A*). The most important upgrade to the transportation infrastructure and a typical feature of urban sprawl has been the addition of a "ring-road" highway. In this case, it is the Kilpatrick Turnpike (*B*). Placemark *C* illustrates a common phenomenon associated with sprawl, the conversion of prime agricultural lands into urban space.
>
> In the last fifteen years, downtown Oklahoma City has been rejuvenated by large amounts of public and private investment. These investments have aided the forces of centralization. View the placemarks included in the *Centralization Example* folder. Public projects include new sports and entertainment facilities (*1*). This investment is at least partially responsible for the relocation of a professional basketball franchise to Oklahoma City. The Bricktown Canal (*2*) was added as a water feature not unlike San Antonio's Riverwalk. A plethora of development has occurred along the canal in the last decade. Perhaps most importantly, there has been significant new residential construction (*3*) in the downtown area associated with these projects. This in-filling increases population density, adds to the local tax base, and rejuvenates blighted areas.

Turning it in:

Your instructor will provide you with an explanation of how to submit your results from this assignment. Two possibilities include e-mailing a .kmz file you create from your *You Map It!* folder or submitting a typed list of the stadiums, their city and latitude/longitude locations, and their attributes and liabilities along with screen shots from Google Earth™.

To create a .kmz from your *You Map It!* folder, simply click once on the You Map It! folder to highlight it, then go to File, Save Place As…, and save it in an appropriate location on your computer.

Name: ______________________________

Date: ______________________________

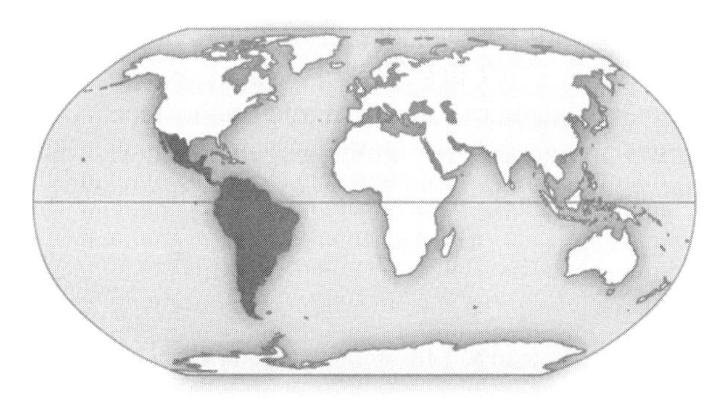

Chapter 4: Latin America

Latin America is a region that geographically includes Middle America (Mexico and Central America) and South America. Romance languages are dominant in this region with most inhabitants of the region speaking Spanish, Portuguese, or to a lesser degree, French. The population of the region is approximately 600 million. Brazil and Mexico are the most populous countries and Mexico City and São Paulo are the most populous cities. Brazil is also the largest country in the region.

Latin America is a region with plentiful natural resources and numerous opportunities for development and prosperity in coming years. The recent award of the 2014 FIFA World Cup and 2016 summer Olympics to Brazil demonstrates this. The most discussed resource issue in the region is the conservation of the Amazon rain forest. Expansion of transportation networks has enabled the exploitation of this resource but these roads also provide the opportunity for economic development. The Amazon is not the only part of the region undergoing dramatic land cover change as large-scale agricultural development schemes are being implemented across the region. As a result this region is increasingly a major player on the stage of global agricultural trade.

The region is home to a highly urbanized population and boasts a number of megacities and/or primate cities. Within these expansive urban environments, there are discrepancies in affluence apparent in the locations, extent, and distribution of urban forms. Many people living in poverty have informal housing that is situated in areas prone to natural disasters.

One realm of Latin America where its inhabitants find common ground is the football pitch. That means soccer field for our North American readers. Fútbol is the most popular sport throughout this region and the fields of play are omnipresent. With the next World Cup in Brazil, the fever will grow even more.

While the arrival of an international sporting event to the region will display Brazil's mercurial rise, challenges remain in the region. For example, the production of coca leaf for cocaine manufacturing presents difficulties for governments both within and external to the region. Local governments face difficult choices in their attempts to stem the drug trade without increasing poverty in rural communities.

As previously mentioned, roads can be an important driver of economic growth in the region. In the case of Latin America, topographic barriers in the forms of mountains, rivers, and dense forests create barriers for region-spanning transportation corridors. Examining these constraints helps one understand the enhanced importance of the maritime corridor through the region. The upcoming changes in the Panama Canal Zone will further cement this feature as one of the globe's key economic choke points.

Download EncounterWRG_ch04_LatinAmerica.kmz from ***www.mygeoscienceplace.com*** *and open in Google Earth™.*

Exploration 4.1: LATIN AMERICA ENVIRONMENT

Forest resources in Latin America have been under pressure for many decades now. While global awareness of deforestation has helped slow the rate of forest loss in some locations, a landscape that has undergone dramatic change is evident when viewing the region in Google Earth. Click the *BR 230 East* placemark and follow the linear feature westward until you reach the *BR 230 West* placemark. Zoom in and out as necessary to examine the Brazilian countryside along this feature. Zoom to the circular anthropogenic features by clicking the placemark titled *strange circles*. Think about what these strange features might be. Is that smoke in the southeast corner of the area containing the features?

Exploration 4.1: MULTIPLE CHOICE

1. After studying the linear feature that connects *BR 230 East* and *BR 230 West*, what is the most likely explanation of what you are examining?

A. an oil and gas pipeline
B. the Transamazonia Railway
C. a road/highway
D. a long-lot farm
E. a fireline

2. Utilizing your knowledge of the region and any additional research you have completed, which of the following words or phrases are most likely associated with the circular features of the *strange circles* placemark?

A. charcoal
B. fish farm
C. candomblé
D. cattle fattening
E. mahogany

Now look at the two pairs of images from South American forests contained in the *Rondonia, Brazil* and *Iguazu, South America* folders. Examining each of the sites individually, toggle the images on and off by checking and un-checking the boxes next to the images to see the change that occurred over a three-decade period. Then examine the patterns surrounding the *Tierras Bajas* placemark.

3. Which of the following statements regarding large-scale deforestation in Latin America is *not* supported by the images in and around the *Tierras Bajas* placemark?

A. Deforestation is taking place in low elevation zones in or adjacent to river basins.
B. Deforestation can occur in areas that are well connected by regularly-spaced road networks.
C. Deforestation has occurred with highly planned/spatially organized development schemes.
D. Agricultural activities that have taken the place of Latin American forests can have very different spatial signatures.
E. Deforestation is a problem largely contained in Brazil.

Turn on the national-level deforestation data by checking the box next to *Forest Cover Change, 1990-2000*. Countries experiencing net forest gain are illustrated with green polygons, while countries exhibiting net forest loss are represented by green polygons. The more a polygon is extruded (the taller the polygon), the greater its value. Maneuver the globe so you can see the varied values for countries in Latin America and around the world.

4. The total amount of deforestation in Latin America is greatest in Brazil, but this data considers the size of the country in the equation to yield a percentage of deforestation. What Latin American country lost the greatest percentage of its forest from 1990-2000?

A. Bolivia
B. Brazil
C. Chile
D. Costa Rica
E. Honduras

5. Compare the deforestation rates that occurred around the world during this period. Which statement is most accurate?

A. No world region showed signs of increasing their forest coverage.
B. According to these statistics, China had one of the 10 greatest increases in forest coverage.
C. According to these statistics, Algeria had one of the 10 greatest increases in forest coverage.
D. Indonesia's declining forests are not significant when one considers the large size of the count and views the percentage loss.
E. Sub-Saharan Africa has shown significant growth in forest coverage in the period illustrated.

Exploration 4.1: SHORT ESSAY

1. Examine the deforestation data, clicking on each country to see where it ranked during the study period. During this time, how would you generalize the deforestation situation in Latin America? What country was the exception to the general trend in Latin America at this time? Can you find some information that would explain this exception? How would you rank Latin America in comparison to other regions at the time?

____
____
____

2. Concerning the forests of the region, what policies could be implemented that would promote long-term economic development while simultaneously working to conserve forest resources?

____
____
____

Remember, when you have completed the questions associated with this exploration, uncheck the 4.1 ENVIRONMENT folder.

Exploration 4.2: LATIN AMERICA POPULATION

Let's examine some Latin American megacities (those with a population exceeding 10 million) and primate cities (those that dominate the economic, political, and cultural spheres of their respective country) with Google Earth. Open the POPULATION folder and view the three cities placemarked. After you examine the cities, it may be helpful for you to view the population density overlay. Double-click the *Population Density* folder, and then check the box to turn it on. Be sure to turn off the overlay when you have completed your study of population density. Now turn on the *La Paz district*, the *San Salvador district*, and the *Medellin district* polygons. Zoom to the respective districts and study the visible differences between the areas of interest.

Exploration 4.2: MULTIPLE CHOICE

1. Which of the following cities is most likely a primate city? (Based on the appearance of the city in comparison to other urban centers within each respective country)

A. La Paz, Bolivia
B. San Salvador, El Salvador
C. Medellin, Colombia
D. La Paz, San Salvador, and Medellin are all primate cities.
E. Neither La Paz, nor San Salvador, nor Medellin is a primate city.

2. Examine the districts outlined in yellow in La Paz, San Salvador, and Medellin. In which city would you classify the highlighted district as a potential squatter settlement?

A. La Paz, Bolivia
B. San Salvador, El Salvador
C. Medellin, Colombia
D. None of the districts appear to be squatter settlements
E. All of the districts appear to be squatter settlements

Now double-click the *Caracas high-income housing* placemark and study the neighborhood. What attributes suggest it is a high-income neighborhood? Now double-click the *Caracas low-income housing* placemark and note the differences from the previous scene. Unlike what we see in the low-income neighborhood of Caracas, there are urban landscapes in Latin America that are highly organized and exhibit a high degree of planning. For example, go to the placemark for Brasília and study the image. Be sure to zoom in and look at some of the links to *Panoramio Photos* found the *Primary Database* of Google Earth.

3. Which of the following is *not* a logical conclusion based on the neighborhood comparison in Caracas, Venezuela?

A. The average home size in the low-income neighborhood is smaller.
B. The residents in the high-income neighborhood face a greater risk of their homes being flooded.
C. Transportation infrastructure is more apparent in the high-income neighborhood.
D. The lot size of residences in the high-income housing is much larger.
E. There is more privacy in the high-income housing.

Master planning and spatial organization is not only evident at the level of national governments or wealthy gated communities. A large portion of Lima, Peru began as a shantytown that later evolved into a developed community. Go to the *Lima, Peru* placemark and search for a pattern that suggests a high level of spatial organization of low-income households. You will need to zoom in to locate the area.

4. What is the name of the former shantytown in Lima, Peru that has become highly organized (e.g. spatial organization evident via Google Earth)?

   A. Dharavi
   B. Los Hoovervíllas
   C. Villa El Salvador
   D. Colonias
   E. Favela

Be sure the *Street View* layer within the *Primary Database* is turned on. Click the link to go to the Plaza de la Constitución in Mexico City and then select "show full screen." Tour around the square by clicking on the small camera icons.

5. Although this plaza predates European conquest, what is the most obvious reflection of colonial European influence from the street view?

   A. the skyscraper
   B. the presence of taxi cabs
   C. the large Spanish flag
   D. vehicles driving on the left side of the road
   E. the cathedral

Now we will zoom out and examine Mexico City from two different eye altitudes. First, double-click the *Mexico City from 80 kilometers* placemark, and then double-click the *Mexico City from 10 kilometers* placemark. Now go to your current city of residence, examine it from 80 and 10 kilometers above the surface and note the differences in comparison to Mexico City.

Exploration 4.2: SHORT ESSAY

1. What aspects of the aerial view and Panoramio photographs of Brasília suggest that this was a master-planned community? What do you think were the goals of the planning body from functional and symbolic perspectives?

   ____
   ____
   ____

2. Describe the differences between Mexico City and your current city of residence based on their appearances from 10 and 80 kilometer eye levels. For example, what is the approximate diameter of the respective cities/agglomerations? What limitations has topography imposed on the development of each location? What transportation infrastructure is evident in each city?

   ____
   ____
   ____

Exploration 4.3: LATIN AMERICA CULTURE

The 2014 World Cup is coming to Brazil and Rio de Janeiro will host the 2016 Summer Olympics. Rio has been a popular tourist destination for many years. The beauty of the city is apparent from the 360-degree view from Sugar Loaf Mountain. Double-click the *Sugar Loaf Mountain* link to see the view. Now do the same thing with the *Ipanema* link.

Exploration 4.3: MULTIPLE CHOICE

1. Which of the following statements would you disagree with after examining the view from Sugar Loaf Mountain?

   A. Rio is a sub-tropical location.
   B. Rio has multiple assets for tourist-based industry.
   C. Rio lacks a harbor.
   D. Rio is densely populated.
   E. Rio is located in an area of high relief.

2. What can you ascertain from the image of Ipanema Beach?

   A. Latin Americans and the tourists in this image require much more personal space than the typical North American
   B. the beach is a hazardous location that the population avoids
   C. everyone brings their own personal umbrella to the beach
   D. there are strict morality laws in Rio that prohibit women from exposing their skin
   E. property values in the immediate vicinity are likely quite high

The focal point of the 2014 World Cup will be the Estádio de Maracanã. This famous fútbol stadium previously hosted a World Cup final in 1950. Double-click the *Maracanã* placemark. Make sure you have the *3-D Buildings* layer turned on in the *Primary Database*. Examine the stadium and the surrounding area. The Brazilian national team plays many of its matches here.

2. Which of the following assets for hosting a World Cup final is evident from the imagery of Estádio de Maracanã?

   A. a large, modern stadium
   B. diverse and numerous mass transit opportunities
   C. a central location
   D. a vibrant, green pitch
   E. Estádio de Maracanã has all of the attributes listed

The countries of South America compete against one another through the Conmebol Confederation. These games have added importance when the countries are attempting to qualify for the World Cup. While Brazil and Argentina are the traditional powers of South America, there are certain venues that all teams dread. Two of those venues are contained in the *Stadia of Dread* folder. Examine the stadiums and also the environments in which they are located.

4. What is the common attribute that the two sites in the *Stadia of Dread* folder possess on a continual basis that would make it more difficult for visiting teams to win?

A. high temperatures
B. sunny weather
C. airport noise
D. thin air
E. pollution from adjacent super highways

5. While soccer represents a unifying cultural trait for Latin America, there is common cultural ground beyond the soccer fields. Think of the sites you have visited and the images you have seen in this exploration. What element of cultural geography would be most difficult to view evidence of from the imagery available via Google Earth and its associated applications (e.g., Panoramio Photos)?

A. religion
B. level of development
C. predominant language
D. urban morphology
E. government type

Exploration 4.3: SHORT ESSAY

1. Examine a language map of Latin America. Describe the relationship between the location and distribution of indigenous languages and facets of the natural environment viewable in Google Earth.

____

____

____

2. The European imprint on Latin America is unmistakable, from its soccer fields to the Catholic churches that anchor the Colonial Era squares. Locate a cultural landscape feature that would not be the result of European influence. Explain your choice and provide the coordinates of the site.

____

____

____

Exploration 4.4: LATIN AMERICA GEOPOLITICS

Click the *La Asunta Chica* placemark. The United Nations has identified this as a coca field. The hyperlinked document contains a wealth of interesting information on coca cultivation and the coca trade. Now look at the National Geographic placemark titled *Cocaine Country*. Click the link and explore the pictures and text associated with the story.

Exploration 4.4: MULTIPLE CHOICE

1. The *Asunta Chica* placemark illustrates that remotely sensed imagery at this resolution

   A. can be an effective tool for identification of potential coca fields.
   B. can positively identify persons involved in drug trafficking.
   C. can be used by law enforcement to identify if armed guards are at a site.
   D. can show exactly what the land cover is at this moment.
   E. is all that's necessary to convict someone of drug trafficking.

Changing territorial boundaries can be a source of conflict between various political entities. Therefore, historical maps can be useful in helping to understand the roots of these tensions. Check the box next to the *Columbia 1840* historical map from the David Rumsey Collection and then double-click the folder to zoom to the appropriate scale. Be sure the *Borders and Labels* layer in the *Primary Database* is turned on. Adjust the transparency of the *Columbia 1840* map by using the transparency slider located at the bottom of the Places pane. Find a level of transparency where you can see the historical map and contemporary boundaries (in yellow).

2. The *Columbia 1840* historical map best illustrates that

   A. cartographic technology has not changed since 1840.
   B. Ecuador was the largest country in the northern tier of South America in 1840.
   C. the northern tier of South America was politically more divided on a national level in 1840.
   D. the northern tier of South America was politically more united on a national level in 1840.
   E. the boundary between Colombia and Brazil has remained static for more than 170 years.

Notice that the coastlines of the *Columbia 1840* historical map and the coastlines depicted by Google Earth do not match in many locations. This is likely the result of cartographic errors and/or differences in the methods used to depict the spherical Earth (map projections), rather than wholesale changes in the locations of the coastlines.

3. What is a feature on the map that would be more likely to have real-world changes in location?

   A. mountain ranges
   B. the Isthmus of Panama
   C. the cities of Bogota and Maracaibo
   D. the channels of sediment-laden streams
   E. the location of Lake Maracaibo

Border conflicts with its neighbors and the ongoing pressures of the coca trade and the Revolutionary Armed Forces of Colombia (FARC) have influenced the behaviors of the Colombian government. Do you think these issues would make Colombia more or less democratic? Open the *Freedom House* link, and examine the data from the most current year available.

4. Based upon knowledge from the National Geographic and United Nations sources along with information in your text, what are two countries that are at the heart of coca-growing and cocaine production?

   A. Brazil and Argentina
   B. Ecuador and Chile
   C. Bolivia and Colombia
   D. Venezuela and Panama
   E. Bolivia and Peru

5. Do any of the following countries in Latin America have the same levels of political rights and civil liberties (as defined by the scores on Freedom House) as the United States and Canada?

   A. Argentina
   B. Brazil
   C. Chile
   D. Venezuela
   E. No countries in Latin America have the same levels of political rights and civil liberties

Exploration 4.4: SHORT ESSAY

1. Based upon the fact that coca thrives on slopes between 500 and 1500 meters above sea level and it is more likely to find narcotics cultivation in less-developed countries, what Latin American countries are most likely to contain coca fields? Use Google Earth to examine terrain/elevation and development indicators from your text or the United Nations Human Development Index. What other information would be useful in identifying locations more likely to contain coca fields?

   ____
   ____
   ____

2. Examine the political rights and civil liberties scores around the region and peruse the overviews of freedom on the *Freedom House* website. Summarize the status of democracy in the region. Are there trends toward more or less freedom evident?

   ____
   ____
   ____

When you have completed the questions for Exploration 4.4, turn off the 4.4 GEOPOLITICS folder.

Exploration 4.5: LATIN AMERICA ECONOMY AND DEVELOPMENT

Expand the *Panama Canal Expansion Project* folder. Click the title of the folder, and then click the link to the Panama Canal Authority and examine the document. Be sure you read pages 8 and 9 of the PDF. Now examine the *Atlantic Locks* and *Pacific Locks* placemarks. These overlays highlight the proposed and accepted solutions discussed in the document you have already read. Remember that you can use the overlay transparency tool to adjust the transparency of the overlays.

Exploration 4.5: MULTIPLE CHOICE

1. What is the basic *engineering* challenge that the Panama Canal addresses?

A. Water moves rapidly from east to west as the Earth revolves.
B. Goods must be moved around the world as quickly as possible.
C. Ships traversing the canal must be lifted over the continental divide.
D. Middle and South America must be held together or they will drift apart.
E. Ships must be unloaded at the lower Pacific side and reloaded at the higher Caribbean side.

2. The largest ships in the world cannot fit in the Panama Canal; which of the following is the solution supported by the majority of Panamanians in a recent (2009) referendum?

A. Build new larger locks.
B. Transport the cargo by air.
C. Close the canal in 2025.
D. Build a transcontinental highway to move the goods.
E. Delay action and revote in 2025.

Be sure the *3D Buildings* folder is turned on in the *Primary Database*, and then click the *Miraflores Locks* folder. Click the links to the historical photos to get a feel for the size of the canal. The Panama Canal can be described as a global choke point. Research the term "choke point" on the web. Then open the *Panama Canal webcams* link and explore the activity that is taking place.

3. The Panama Canal is a global choke point. Which of the following locations is the most similar type choke point?

A. Malacca Strait
B. Strait of Gibraltar
C. Khyber Pass
D. Erie Canal
E. Suez Canal

The Pan-American Highway is a network of highways and roads that stretches from Alaska to Chile and Argentina. However, this is one major gap where no road exists. The Darién Gap, in portions of Panama and Colombia is one of the world's biodiversity hot spots. It's also dangerous because of armed guerillas operating in the area. Therefore, building a road would be expensive, dangerous, and damaging to the environment. Click the *Pan-American Highway* placemarks and then zoom in to see if you can find any signs of human activity in the Darién Gap.

4. One of the *Panama Canal webcams* displays the Centennial Bridge, an important link on the Pan-American Highway. Which of the following regional trade pacts would benefit most from the possible completion of the Pan-American Highway?

A. CAFTA
B. NAFTA
C. ECOWAS
D. MERCOSUR
E. EU

5. What is the approximate straight-line distance of the uncompleted gap in the Pan-American Highway?

   A. 10 km
   B. 75 km
   C. 120 km
   D. 250 km
   E. 5,000 km

Exploration 4.5: SHORT ANSWER

1. Determine the approximate mileage of a ship traveling from New York City to Long Beach, California via the Panama Canal versus going around Cape Horn. What is the difference in distance? If your ship was larger than a Panamax vessel, what are two options that you could pursue to get your goods to Long Beach? In your description of each alternative, think and write about the different costs you might incur with each method.

   ____
   ____
   ____

2. Do you support the idea of completing the unfinished link of the Pan-American Highway? What are the benefits and drawbacks to completing the Pan-American Highway through the Darién Gap?

   ____
   ____
   ____

YOU MAP IT! World Cup 2014

You have visited the home of the 2014 FIFA World Cup final earlier in this exercise. However, to stage a successful World Cup, it is necessary for the host country to have numerous world-class stadiums. The 12 host cities for the 2014 World Cup include Rio (home of the Maracanã), Belo Horizonte, Natal, and Brasília. These four stadiums have been mapped for you. To see them, turn on the *You Map It!* folder. Your job is to identify and find the additional eight stadiums that will host matches. When you find them, add a soccer player placemark and a title of the city. Be sure your new placemarks are created in the *You Map It!* folder. Scale the size of the city title to the relative size of the stadium. The Maracanã holds close to 100,000 persons (although a crowd of more than 200,000 watched the 1950 final there). On the other hand, the stadium in Brasília only holds 40,000 and will likely be expanded in the coming years. Use these stadiums as the high and low ends for your capacity scale.

Change the color of the city text to red, yellow, or green. Red lettering suggests that the stadium/site has a long way to go to be ready for the World Cup. For example, the stadium may be too small, the field could be in poor condition, and/or there may be no mass transit (highways and rail lines) nearby. Yellow lettering suggests that one or two of these types of problems are apparent, but that the stadium/site is marginally prepared for the event. Green lettering indicates that your analysis has concluded that the site is more or less ready for the event. I would encourage you to look at Panoramio photos and web links that can provide information on stadium capacity and condition to help with your assessments. In the description box of your placemark, comment about three to five elements that stand out in your analysis.

Before you start this mapping project, be sure to gain a comparative context for your commentary by viewing the collection of stadiums from the 2006 World Cup by following the link in the *You Map It!* folder.

Turning it in:

Your instructor will provide you with an explanation of how to submit your results from this assignment. Two possibilities include e-mailing a .kmz file you create from your *You Map It!* folder or submitting a typed list of the stadiums, their city and latitude/longitude locations, and their attributes and liabilities along with screen shots from Google Earth.

To create a .kmz from your *You Map It!* folder, simply click once on the You Map It! folder to highlight it, then go to File, Save Place As…, and save it in an appropriate location on your computer.

Name: ______________________________

Date: ______________________________

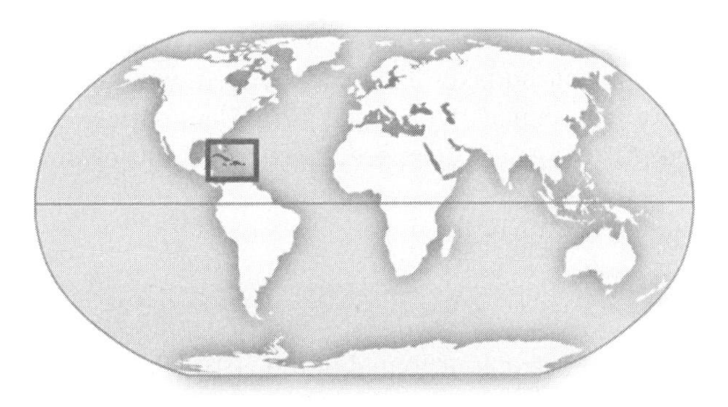

Chapter 5: Caribbean

The Caribbean region consists of the Caribbean Sea, its islands, and some of the surrounding coastal areas of Latin America. It is a culturally diverse region, a product of indigenous, European colonial and African influences that were historically merged in the context of plantation agriculture.

The population of the region is concentrated on the four larger islands of the Greater Antilles. Cuba has the largest population, with approximately 12 million inhabitants and Havana is the region's largest city with a population that approaches four million.

While Cuba is the largest country in physical size as well, Cuba is not familiar to many people of the world. This is at least partially the result of a US embargo against Cuba that has existed since the early 1960's.

The region is situated atop the Caribbean tectonic plate. The island arcs of volcanic origin delineate the margins of the plate on its eastern side. Volcanic activity devastated the island of Montserrat in the middle and late 1990's. The active tectonic situation is also evident by the occurrence of earthquakes, such as the devastating temblor that occurred near Port-au-Prince in 2010.

While the challenges presented by the environment are formidable, the Caribbean region's environment is one of its greatest assets. Millions of tourists flock to the idyllic beaches and azure waters of the Caribbean. Therefore, tourism is a central component of the regional economy.

However, tourists don't come to the Caribbean for the beaches alone. World Heritage sites are locations that have been recognized as globally significant for their natural and/or cultural attributes. In the Caribbean region there are numerous World Heritage sites with the majority recognized as locations of cultural importance. As one explores these sites, the imprint of European colonialism in the region is conspicuous.

European powers have not been alone in exercising their influence in the region. In the early 19th century, the US declared that it would not tolerate further efforts by the Europeans to colonize lands in the Caribbean. This subsequently set the stage for the US to become the preeminent power in the region.

While the US is unquestionably considered the most influential military power in the region, the European legacy is still quite evident. Formal relationships between a number of Caribbean political entities and their former European masters remain to this day.

Download EncounterWRG_ch05_Caribbean.kmz from ***www.mygeoscienceplace.com*** *and open in Google Earth™.*

Exploration 5.1: CARIBBEAN ENVIRONMENT

Examine the major plate boundaries of the Caribbean region via the *Plate Boundaries* folder. Plate boundaries can be generally subdivided into three categories: convergent (plates coming together), divergent (plates moving apart), and transform (plates sliding by one another). Volcanoes and earthquakes can be associated with any of the boundary types. However, earthquakes are most commonly associated with transform boundaries, while volcanoes are most commonly associated with convergent boundaries. As you can see, there is a significant plate boundary near the island of Hispaniola.

Exploration 5.1: MULTIPLE CHOICE

1. What statement most accurately portrays the tectonic realities that contributed to the 2010 Haitian earthquake? Be sure to evaluate the statement in its entirety for accuracy.

A. The eastern flank of the Caribbean Sea contains a 1000-km subduction zone.
B. There is a small transform boundary due east of Port-of-Spain.
C. There is a convergent boundary south of Hispaniola.
D. There are diverging plates west of Hispaniola.
E. There is a transform boundary just north of Hispaniola.

The convergent boundary in the vicinity of the Lesser Antilles represents the collision of the Caribbean and South and North American plates (the boundary between the North American and South American plates has not been clearly defined). As the North and/or South American plate(s) dive under the Caribbean plate we have a subduction zone that contributes to volcanic activity in the Lesser Antilles. There are more than one dozen active volcanoes in the Lesser Antilles. Perhaps the most notable is the Soufriere Hills Volcano on the island of Montserrat. Click the *Soufriere Hills* placemark and then click the link to view images of the volcano captured from the International Space Station. Read the text associated with the images. Now click the *Plymouth* placemark and examine the landscape. As a result of the eruption, Plymouth is no longer a viable capital city for this British overseas territory. Be sure the *Terrain* layer in the *Primary Database* is turned on and explore the island. Think about why the northern part of the island appears to be the location of the majority of the remaining population.

2. What was the primary volcanic hazard that affected the towns of Plymouth and Richmond Hill?

A. lava flows
B. volcanic spines
C. volcano-induced earthquakes
D. pyroclastic flow
E. ash fall

3. In your estimation, what percentage of *Plymouth* was inhabited at the time this image was captured?

A. 0
B. 30
C. 50
D. 70
E. 100

4. Why is the northern part of Montserrat the location of the majority of the remaining population?

   A. It has a large river that divides it from the southern part of the island and would stop any lava flows.
   B. It is sheltered somewhat from volcanic activity by a ridge of higher ground in the middle of the country.
   C. It is the part of the island with beach resorts.
   D. It is where the large agricultural plantations are located.
   E. The northern population centers are more than 50 kilometers from the volcanic activity.

Use Google Earth™ to locate the following cities in the Lesser Antilles: The Valley on Anguilla, Marigot on Saint Martin, Le Morne Rouge on Martinique, St. Georges on Grenada, and Codrington on Barbuda. Evaluate the local topography and determine which of these cities appears to be in the most hazardous position if a nearby volcano erupted.

5. What city in the Lesser Antilles appears to be at the greatest risk from a volcanic eruption?

   A. The Valley
   B. Marigot
   C. St. Georges
   D. Codrington
   E. Le Morne Rouge

Exploration 5.1: SHORT ESSAY

1. While earthquakes and volcanoes are a clear and present danger to many inhabitants of the Caribbean, there are some locations that are particularly susceptible to devastation when a tectonic event occurs. Think of the 2010 earthquake that struck Haiti. It was a magnitude 7.0 earthquake. Similar-sized earthquakes struck the United States in the 1990s with only a fraction of the lives lost. Explain the factors that contributed to make the Haitian event so much more devastating.

   ____
   ____
   ____

2. In Exploration 1 you have examined the seismic hazards of the Caribbean region. However, the impacts from tropical cyclones are usually more significant on an annual basis. This region is one of the highest-frequency regions for the development of tropical cyclones. Why is the Caribbean prone to what are called hurricanes in this region? Secondly, why is this region disproportionately affected by hurricanes beyond the fact that they are more frequent here?

   ____
   ____
   ____

Exploration 5.2: CARIBBEAN POPULATION

The US embargo on Cuba has been in place for half a century and its impacts are evident at the street-level. If you are not sure what is meant by the term embargo, research the phrase now. Open the *Havana Scenes* folder and examine the five street views of Havana. Now examine the scenes included in the *San Juan Scenes* folder. Do the scenes from San Juan seem more or less distinctive than those from Havana? Which place looks like a better place to live? Which looks most interesting to visit?

Exploration 5.2: MULTIPLE CHOICE

1. Upon completion of your examination of the *Havana Scenes*, which of the following statements would best reflect the repercussions of a trade embargo?

A. The fashion (e.g. hair styles, shoes, and clothing) evident in the images appears to be strictly pre-1960's.
B. There is a preponderance of pre-1960's automobiles.
C. The technology is not present to build multi-story buildings.
D. It is clear that Havana is a high-technology hub.
E. Havana does not have electricity.

2. Compare the scenes of Cuba's capital city with your knowledge of the following capital cities: Washington D.C., Paris, and Tokyo. Havana's economic development appears to be

A. similar to that of the aforementioned capital cities.
B. generally more advanced than the aforementioned capital cities.
C. generally less advanced than the aforementioned capital cities.
D. something that cannot be judged from appearances.
E. more advanced in terms of tourism, but less advanced in terms of transportation infrastructure.

3. In terms of general levels of development and cultural distinction as evidenced by a unique sense of place, what statements would you agree with following your assessment of the *San Juan* and *Havana* scenes?

A. Havana is more developed and more culturally distinct.
B. San Juan is more developed and more culturally distinct.
C. Havana is more developed and San Juan is more culturally distinct.
D. San Juan is more developed and Havana is more culturally distinct.
E. Havana and San Juan appear to have very similar levels of development and cultural distinction.

4. Examine the *San Juan 4* scene. Use your detective skills to answer the following question. This location is

A. the site of a colonial-era French fort.
B. the site of Puerto Rico's first national capital.
C. the Institution of Kite Dynamics.
D. an Operation Bootstrap headquarters.
E. a US Park Service National Historic Site.

Now that we have looked at Havana and San Juan at street-level, let's take a different perspective. Click the *Havana Scenes* folder and you will see the Havana metropolitan area from an eye altitude of approximately 15 kilometers. Do the same thing with the *San Juan Scenes* folder and compare the two cities. Zoom in and out as necessary. Keep in mind that the metropolitan population of Havana (approximately 2.5 million) is about five times that of San Juan.

5. After studying the Havana and San Juan urban environments from a bird's-eye perspective, select the most accurate statement.

A. Havana and San Juan are nearly mirror images with respect to population density, transportation infrastructure, and open space in and around residential areas.
B. San Juan appears to have a higher population density and a larger rail hub at its port, while Havana has more large highways and more open space in and around residential areas of the city.
C. Havana appears to have a higher population density and a larger rail hub at its port, while San Juan has more large highways and more open space in and around residential areas of the city.
D. San Juan appears to have a higher population density, more large highways, a larger rail hub at its port, and more open space in and around residential areas of the city.
E. Havana appears to have a higher population density, more large highways, a larger rail hub at its port, and more open space in and around residential areas of the city.

Exploration 5.2: SHORT ESSAY

1. You probably noticed that there were far fewer motor vehicles on the roads of Havana than most similarly-sized cities. What could explain the lack of vehicular traffic? Be sure to provide and develop at least two possibilities in your response.

____
____
____

2. After viewing the city images of Havana, are you more or less interested in traveling there? Identify at least three factors that make Havana look appealing and at least three that make it seem undesirable.

____
____
____

Exploration 5.3: CARIBBEAN CULTURE

Within the *Culture* folder there is a *World Heritage Sites in the Caribbean* folder containing a link to World Heritage information from the World Heritage Committee. The 12 placemarks within this folder will lead you to the culturally-oriented World Heritage sites of the Caribbean. To complete the questions for this exploration, you will utilize Google Earth™ in conjunction with information available from the World Heritage Committee. Use the "World Heritage Info" link in the *World Heritage Sites in the Caribbean* folder to access this information. By clicking "The List" tab, information, maps, and photos for individual sites can be obtained. Two of these sites can be viewed from a 360° city scene.

Exploration 5.3: MULTIPLE CHOICE

1. Which of the following statements does *not* accurately reflect the mission of World Heritage?

   A. Promote international cooperation to protect heritage resources
   B. Protect heritage resources from exploitation by local populations
   C. Encourage member states to develop management plans to protect resources
   D. Encourage member states to nominate their sites for inclusion on the World Heritage List
   E. Encourage international cooperation in the conservation of our world's cultural and natural heritage

2. Click the link for the Viñales Valley and explore the area. Complete the following statement based upon your examination of the landscape and information on the site available from the World Heritage Committee via the "World Heritage Info" link. The Viñales Valley has been recognized by UNESCO as

   A. a unique urban landscape that combines traditional architecture with Soviet-era architecture.
   B. a site with significant archaeological remains of earlier Islamic culture.
   C. the first tourist site in the Caribbean.
   D. a unique cultural landscape with traditional methods of agriculture focused on tobacco production.
   E. a unique cultural landscape that illustrates plantation agriculture focused on cotton production.

Look closely at the *Old Havana* scene and explore the products available in the street market. One of the two principle agricultural exports of Cuba during its more prosperous trading period of the late 19th century figures prominently in this image. Examine the remaining culturally-oriented World Heritage sites in the region by clicking the respective placemarks and reading their descriptions at the World Heritage website. After exploring and reading about these sites, does it seem like the recognized values of the sites are often directly or indirectly related to the processes of and responses to European colonialism of the Caribbean?

3. The agricultural product spread on the ground in the *Old Havana* scene has been one of the most important commodities of Cuba since the time of colonization. It is

   A. sweet potatoes.
   B. limes.
   C. cotton.
   D. tobacco.
   E. sugar cane.

4. Which of the following sites, based on the brief descriptions from "The List," represents the most acute example of early globalization? In this case we are specifically thinking about slave labor from one region in the world (Africa) being directed by controlling powers from another world region (Europe) building something in a third region (the Caribbean).

   A. Brimstone Hill Fortress
   B. Colonial City of Santo Domingo
   C. Historic Inner City of Paramaribo
   D. Historic Area of Willemstad
   E. San Pedro de la Roca Castle

5. How many of the 12 cultural World Heritage sites in the Caribbean have recognized values that are directly *or indirectly* related to the processes of, and responses to European colonialism?

A. none of the sites are related to European colonialism
B. a few of the sites are related to European colonialism
C. half of the sites are related to European colonialism
D. most of the sites are related to European colonialism
E. all of the sites are related to European colonialism

Exploration 5.3: SHORT ESSAY

1. Utilizing the information that you have acquired from this Exploration along with your textbook and any additional outside resources, provide a brief synopsis of European colonial involvement in the Caribbean. What countries were active in the region? Do any of these countries still have formal political relationships with their former colonies?

____

____

____

2. Explore the World Heritage website in greater detail. Identify the sites in the Caribbean that have been protected for their outstanding physical attributes. Provide the names, locations, and a brief description of the outstanding attributes for which they have been recognized.

____

____

____

Exploration 5.4: CARIBBEAN GEOPOLITICS

A keystone event of the US military dominance in the region and also a moment when the Cold War nearly escalated to a nuclear conflict is known as the Cuban Missile Crisis. This event was triggered by the Soviet Union placing nuclear missiles in Cuba during the fall of 1962. The US demanded that the missiles be removed. How had the US verified the presence of these sites? They had utilized aerial photography captured by reconnaissance aircraft. Expand the *Cuban Missile Sites* folder and you will see five of the sites involved in the crisis. Half a century after the confrontation is there any visible evidence to suggest these were once nuclear missile launch sites? Examine each site by clicking the respective placemarks. You can see that some sites are being used now for manufacturing or mining, there are storage buildings at another, and two of the sites look to have been abandoned for some time.

Enable the *Panoramio Photos* layer in the *Primary Database*, and then zoom to the *San Cristobal III* and *Sagua la Grande I* sites. Click the small blue boxes to view the photographs from the locations. Use your mind to translate the way features look differently from an aerial versus a ground-based perspective.

Exploration 5.4: MULTIPLE CHOICE

1. Upon exploring the former Cuban Missile Sites, which location is the most remote and appears to have been overgrown with forest?

A. Guanjay I
B. Guanjay II
C. Sagua la Grande I
D. Sagua la Grande II
E. San Cristobal III

2. After viewing the *Panoramio* photos in conjunction with the aerial views from Google Earth™, what unique feature can you correlate between the two perspectives?

A. the access road at Sagua la Grande I
B. the white stone pillar at Sagua la Grande I
C. the gate at San Cristobal III
D. the cattle at San Cristobal III
E. the lighter color of the concrete pads at San Cristobal III

The Cuban Missile Crisis is not the last time the US has flexed its military muscle in the region. Did you realize that the US actually invaded a Caribbean country in the 1980's? Turn on the *Invasion Map* and view the plan for the assault on this Caribbean island. Based upon the features of the map, determine what island country is represented. When you determine the country, research the year and circumstances of the invasion.

3. In what Caribbean country and in what year was there a US invasion?

A. Dominican Republic, 1965
B. Cambodia, 1970
C. Grenada, 1983
D. Cuba, 1988
E. Haiti, 1999

The US military presence in the region remains today. This presence has been welcomed by Caribbean nations in times of crisis, such as the 2010 earthquake in Haiti. Other times, Caribbean populations have had mixed feelings about US military activities in the region. Click the *Vieques* folder. This island was home to US Naval training from 1941 to 2003. The Navy used Vieques for bombing practice and the storage of military and industrial waste. Today, much of the island has been converted to a wildlife refuge. The physical scars from bombing will take a number of years to heal. Examine placemarks *1*, *2*, *3*, *4*, and *5* in the *Vieques* folder.

4. Examine the landscapes around the five numbered placemarks contained in the Vieques folder. Which landscape presents the clearest example of a surface altered by munitions detonation?

A. 1
B. 2
C. 3
D. 4
E. 5

While the US presence in the region is conspicuous, the European influences remain in cultural, social, and economic realms. Click the *European Mystery Site* placemark and study the unique complex of structures at the location. Using Google Earth™ and outside resources if necessary, identify this site. When you have completed this exploration, turn off the *GEOPOLITICS* folder.

5. The *European Mystery Site* is

A. a weapons testing site.
B. the European Cultural Studies School of the Caribbean.
C. a cobalt mining operation.
D. a tropical disease research center.
E. a space launch center.

Exploration 5.4: SHORT ESSAY

1. Research the use of US troops in the Caribbean region since 1950. Identify four situations where the US military was involved and briefly describe the reasons for the application of troops. Be sure to consider disaster relief operations and election stability forces in your response.

____

____

____

2. Determine the purpose of the *European Mystery Site.* Discuss the history of the site and the geographic reasons for its location.

____

____

____

Exploration 5.5: CARIBBEAN ECONOMY AND DEVELOPMENT

White beaches and clear blue waters come to mind for many when they hear the word Caribbean. These resources have proven to be significant assets in the economic development of the region. Many tourists from around the world come to enjoy the pleasant weather and relaxing beaches. Expand and turn on the ECONOMY AND DEVELOPMENT folder. Examine the 360° view of *Orient Beach on St. Martin Island* as an example. Tourism is the number one economic resource for the people of the Virgin Islands. An activity that draws many tourists to the Virgin Islands and the Caribbean region is scuba diving. Fly into the ultra high-resolution gigapan photo, *coral reef scene, British Virgin Islands.* When ocean waters warm up, coral can be bleached and eventually killed. If the world continues to warm, sea-surface temperatures will rise as well. Think about the impacts on tourism alone as more coral becomes diseased and dies.

Exploration 5.5: MULTIPLE CHOICE

1. In reference to warming waters in the Caribbean Sea,

A. scuba tour operators in the Caribbean are pleased because they will be spending less on wetsuits.
B. coral can be damaged and eventually die.
C. the increased evaporation will lower the sea level.
D. the water is becoming less acidic due to increases in dissolved carbon dioxide.
E. coral is evolving to become "super coral" that can survive above sea level.

Tourism can be a controversial form of development. For example, tourists place added burdens on local infrastructures, although they do not individually contribute to the local property tax base. Another problem is capital leakage. This term refers to the fact that most of the profits from tourist enterprises leave the region because many of the owners and operators of the beach resorts and cruise ships reside in Europe, Asia, and the US. Low-paying service jobs are often the only local benefits realized.

Cruises are a popular way for many tourists to explore the Caribbean. Exploring the harbors of the region, it is not difficult to identify a cruise ship. Click on the *Cruise Ship* placemark. You will see numerous features that suggest this is a ship built to carry a large number of people and that entertainment is a primary objective. For example, the ship is multi-story. You can see the ship has at least six levels by looking at the stern. You can also see a large number of life-boats. This suggests that the human capacity of the ship is high.

2. The *cruise ship* has all of the following entertainment features visible except

A. a tennis court.
B. hot tubs.
C. a water slide.
D. at least four pools.
E. miniature golf.

These large cruise ships are eye-catching visitors of Caribbean harbors. Explore the *Harbor A*, *Harbor B*, and *Harbor C* placemarks. Notice the differences in the relative size and facilities in each harbor as well as any cruise ships that are docked in the harbors. When you click the *Montego Bay, Jamaica* folder you will see yet another cruise ship in port. Examine the pier facility and the rest of the area that is visible without zooming in or out. Does this look like a "natural landscape"?

3. After exploring *Harbor A*, *Harbor B*, and *Harbor C*, select the pair of words that identifies the harbor that had no cruise ships and an appropriate explanatory word or phrase.

A. C, military ships only
B. C, no pier to dock
C. A, no cultural or tourist sites nearby
D. B, no pier to dock
E. A, embargo

Within the *Montego Bay, Jamaica* folder, click the *Site A* placemark. Does this site look natural or modified? Humans have clearly altered the landscapes extensively at both these locations. What did these sites look like before the extensive alterations? You do not have to go far. Click *Site B* and you will see a nearby mangrove forest. Mangroves are trees that grow in saline environments and provide crucial protection against coastal erosion as well as habitat for a wide variety of plants and animals. Many of the most popular tourist sites around the Caribbean are located at the locations of former mangrove forests.

Zoom to the *Atlantis Resort* placemark. This is the largest resort in the Caribbean region. In the 2005 imagery you can see that the area is a massive construction site. You can use the time-slider to show historical imagery in Google Earth™. You turn it on and off by clicking the small clock icon in the tool bar. This is but one example of the dramatic landscape-clearing that continues to take place in the region. Atlantis is another example of capital leakage, but it might surprise you to find out where the developers of this mega-resort are located.

4. Research the Atlantis Paradise Island Resort and determine what world region is home to the developer and original owner of the resort.

A. Caribbean
B. North America
C. Sub-Saharan Africa
D. Asia
E. Europe

While humans have been largely responsible for the contemporary distribution of mangrove forests, there are natural factors to consider as well. Examine the periphery of the islands of *St. Lucia* and *Barbados*. What sides of these two islands are the most densely populated? Think about why this is the case. The images actually illustrate the reason. You'll also find out that the mangroves that remain on these two islands prefer the same sides of the island for the same reason. Subsequently, much of the mangrove forests here have been cleared for development.

5. You have examined the coastal peripheries of St. Lucia and Barbados. Why does development *and* mangrove forest prefer the same locations? Identify the explanatory word or phrase.

A. differences in wave energy
B. sea water salinity variations
C. greater sunshine
D. better soils
E. longer days

Exploration 5.5: SHORT ANSWER

1. Cruise ships may be the most recognizable vessels in the harbors of the Caribbean, but what else is out there? Re-examine the harbors of the region and attempt to generally identify three other types of ships. Explain their hypothetical roles in the regional economy.

____
____
____

2. Agricultural exports, despite sugar and bananas leading the way, are a decreasing segment of the regional economy. Based on what you have seen in this Exploration and supplemental information from your textbook, hypothesize as to why this is the case. Develop at least two explanatory threads in your response.

___

___

___

YOU MAP IT! Potential World Heritage Sites

As discussed above, a significant percentage of the Caribbean's World Heritage sites with a cultural emphasis have direct or indirect ties to European colonialism. Utilize the internet and Google Earth™ to identify three sites of universal cultural and/or physical significance in the region. Identify these on Google Maps with placemarks and include a brief description of the attributes that justify your decision. A sample site has been identified in the *YOU MAP IT!* folder. An example description of that site follows.

Blue and John Crow Mountains National Park, Jamaica, 18.14° N latitude, 77.1° W longitude

Blue and John Crow Mountains National Park was established as Jamaica's first national park in 1990. It protects a sizable percentage of Jamaica that still has forest. In fact, the forests contained in the park have some of the highest species diversity in the Caribbean. This diversity is the primary reason this park should be included on the World Heritage List. The second reason is that this rich ecosystem is facing ongoing deforestation as a result of slash and burn agriculture and illegal harvesting of valuable hardwoods. Invasive species are also a problem for the park. These threats could help be addressed by the increased funding that could be derived from World Heritage designation.

Turning it in:

Your instructor will provide you with an explanation of how to submit your results from this assignment. To create a .kmz from your *You Map It!* folder, simply click once on the You Map It! folder to highlight it, then go to File, Save Place As…, and save it in an appropriate location on your computer.

Name: ______________________________

Date: ______________________________

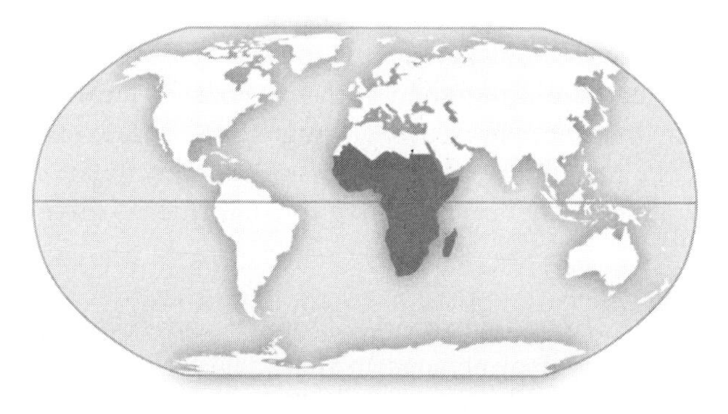

Chapter 6: Sub-Saharan Africa

The Sub-Saharan Africa region refers to the part of the African continent that lies below the Saharan Desert. The Sahel is the transitional zone along the southern margins of the Sahara Desert that separates the more arid region of North Africa from the more humid Sub-Saharan portion of the continent. The Sahel is not only a climatological divide, but also a cultural one.

North of the Sahel, the countries are largely considered part of the Arab world with Arabic the dominant language and Islam the most prevalent religion. Areas south of the Sahel are highly diverse as evidenced by the greatest regional language variation in the world and a complex mix of Christian, Animist, and Muslim religious traditions.

The population of Sub-Saharan Africa is noteworthy for its continued high growth rates. With a regional population that will surpass one billion in the current decade, the most populous country is Nigeria with a population near 160 million. Lagos, Nigeria is the region's largest city with a population approaching 15 million. This region is exceptional in other measures of population such as infant and maternal mortality rate, age distribution, and life expectancy.

The growing population of the region has placed a significant strain on the resource-base of the region. Population pressure has combined with recurrent drought in the Sahel region to create a situation where the grasslands of this transitional zone have been degraded to the point of no longer being viable economic resources. This deterioration of the land is known as desertification and the Sahel represents one of the world's most extensive case studies.

Religious preference is another interesting component of the region's population. This defining cultural element is in a state of flux here more than any other world region. Animist traditions are waning while Christian and Islamic influences are growing. The different traditions provide us with recognizable features on the cultural landscape.

Ethnicity is another cultural component of the region that exhibits heterogeneity. At times, these differences have been used by combatants as cause for persecution. The horrific and ongoing case of Darfur in southwestern Sudan is such a situation. The conflict has displaced millions of people from their homes.

The diversity of Sub-Saharan Africa goes beyond culture. The region is rich in resources. Some feel these resources have fueled a type of neo-colonialism, with rich countries from outside the region plundering the resources with little gain for the people of Africa. Undoubtedly, Sub-Saharan Africa faces some unique challenges in its ongoing economic development.

Download EncounterWRG_ch06_SubSaharanAfrica.kmz from ***www.mygeoscienceplace.com*** *and open in Google Earth™.*

Exploration 6.1: SUB-SAHARAN AFRICA ENVIRONMENT

Africa's Sahel is a transitional region and a region in transition. It represents the middle ground between the driest and wettest parts of Africa and it is also undergoing environmental change. The changes that are taking place can be tied to climatic variation and the individual and collective decisions of humans. Humans have occupied the Sahel for many thousands of years, forging an existence often based on migrating with livestock. We call these migratory people pastoral nomads. The nomads lead their livestock in seasonal migrations based on the availability of forage. The availability of food for the grazers is a function of grazing pressure and precipitation. If there are too many animals using a scarce resource and/or the atmosphere doesn't deliver the moisture to grow the grass, there can be big problems.

Expand the 6.1 ENVIRONMENT folder, and zoom to the *nomad huts* placemark. Zoom in and examine the structures and the landscape. If you look closely you will see individual persons in the image. Look for their shadows. In this image you also see the regularly spaced plants that suggest planting. This is a sorghum field being maintained by these nomads. This strategy of growing grain is often employed by nomads in conjunction with their livestock activities. Examine the *storage* placemark. The round features are granaries where the sorghum and millet are stored. Look closely and you will see two pack animals loaded with sacks of grain. Now zoom to the *nomad 2* placemark. While pursuing a similar adaptive strategy, the image suggests that this group is less transitory than the previous group.

Exploration 6.1: MULTIPLE CHOICE

1. Which of the following most strongly suggests that the people and dwellings visible in the nomad huts placemark are transient?

A. The four-wheel drive vehicles by each dwelling.
B. The line of people entering or exiting the location.
C. The cattle trailers.
D. The construction and materials of their dwellings.
E. The paved roads.

2. Which of the following is *not* a feature that suggests that the nomads 2 settlement is less transitory than the group illustrated in the nomad huts placemark?

A. granaries
B. fences
C. established trees
D. mud construction
E. smaller agglomeration of dwellings

This region has supported low intensity grazing and farming for millennia. Many people in the region are adding increasing numbers of livestock to their agricultural portfolio. However, the increased pressure on the grasslands has contributed to the process of desertification in some cases. You can see a herd of cattle near the *cattle* placemark. Examine the land cover in this photo. Now zoom out to an eye elevation of approximately two kilometers and survey the land cover.

3. Your assessment of the land cover in and around the *cattle* placemark has determined that

A. this is a favorable location for grazing because of the extensive shade provided by trees.
B. this location could sustain many more head of cattle.
C. numerous water resources exist for the herd.
D. there is very limited forage for a large herd of animals.
E. this location is problematic because of the numerous villages herds encounter while grazing.

While human activities are often the acute factor that degrades the productivity of lands in semi-arid and arid regions, drought can make the vegetation more susceptible to this degradation. In areas with soils prone to wind erosion, sand dunes will begin to form and migrate across the landscape. Check out the *desertification* placemark. This image was captured in the vicinity of Lake Chad. This is a large, shallow lake that has varied in size dramatically in last century. Lake Chad is currently just a fraction of the size that it was in the late 1960s. Its diminished size has been attributed to drought as well as diminished vegetation in the region due to overgrazing. Open the link in the *Lake Chad* folder and read the information on Lake Chad provided by NASA. Now expand the Lake Chad folder and open the *UNEP Lake Chad* link to learn more about this dynamic physical feature. Now turn on the year layers in the *Lake Chad* folder one at a time, beginning with *1963*. Observe the changes in the water levels of Lake Chad for each year through the *February 2007* layer.

4. Based upon your study of the imagery sequence and information from the UNEP and NASA links, Lake Chad's water level was lowest in

A. 1963.
B. 1972.
C. 1987.
D. 1990.
E. 2000.

In areas of the Sahel with higher moisture content, woodlands and even forests are present. Expand the *Baban Rafi Forest* folder and click the link to read about the loss of woodlands in the last 40 years. Turn on, and then toggle the *1976* and *2007* layers to visualize the loss of timber in the region.

5. Which of the following is *not* an explanatory variable for the physical changes that have occurred in the Baban Rafi Forest of Niger between 1976 and 2007?

A. a 400 percent decrease in population
B. wood cut for fuelwood purposes
C. selective exploitation of tree species
D. land cleared for agriculture
E. increasing pressures from human settlement

Exploration 6.1: SHORT ESSAY

1. You now know that Lake Chad has decreased in area over the last 40 years as evidenced by satellite imagery. In fact, the level of Lake Chad has fallen and risen throughout contemporary and prehistoric times. Examine the Lake Chad close-up placemark and describe the evidence that this aerial photograph provides for the case that there are short-term variations in the levels of Lake Chad. You will need to be sure that Lake Chad UNEP layers are turned off to see this placemark.

___

___

___

2. Determine what countries contain parts of the current or historic Lake Chad and list these countries. Think about the geopolitical challenges that come along with a shared resource such as Lake Chad. Describe a potential conflict that could arise over this resource and an approach or policy that could be enacted to mitigate this problem.

___

___

___

When you complete Exploration 6.1, uncheck the 6.1 ENVIRONMENT folder.

Exploration 6.2: SUB-SAHARAN AFRICA POPULATION

Population in Sub-Saharan Africa is an issue of great importance. Populations in the region continue to grow rapidly. However, the economic development in many locations in the region is sufficient to support a standard of living that will lead to long and healthy lives. Begin this exploration by expanding the 6.2 POPULATION folder, and turning on the *population growth rate – 2008* folder. Explore the region, and the world for that matter. Click the flags of countries to see their statistical ranking within each category. In Sub-Saharan Africa, there are a number of countries that are among the world's fastest growing. Strangely, however, the southern cone of the region has relatively lower growth rates.

Turn off this folder and turn on *total fertility rate – 2008*. Remember that the total fertility rate represents the number of children an average woman will have in her lifetime. Not surprisingly, these southern cone countries generally have lower fertility rates. Women are having fewer babies, thus the population growth rate is lower. There is one country however that is particularly anomalous. It has a fertility rate that is in the top 15 percent in the world, but it has a negative population growth rate. What could explain this strange relationship?

There is usually a strong correlation between population growth rate and total fertility rate. You can see this relationship in the western part of the Sub-Saharan Africa region. Study the data from the *total fertility rate – 2008* and *population growth rate – 2008* folders. Identify variations in the datasets between individual countries and think about what this means for future population trends in the region. Pay particular attention to Mali, Liberia, and Guinea.

Exploration 6.2: MULTIPLE CHOICE

1. Two countries in Sub-Saharan Africa have negative population growth rates in spite of high total fertility rates. Select the response that includes the country involved and the primary explanatory variable.

A. Swaziland, affluence from diamonds
B. Democratic Republic of Congo, HIV/AIDS
C. Zimbabwe, HIV/AIDS
D. Lesotho, Malaria
E. South Africa, poorest country in region

2. Compare the total fertility rates and the population growth rates of Mali, Liberia, and Guinea. Which of the following statements is *not* correct?

A. An average woman from Guinea will have fewer children than an average woman from Liberia.
B. Mali is currently growing the most quickly on an annual percentage basis.
C. Liberia is growing more quickly than Guinea on an annual percentage basis.
D. The average woman in Mali will have approximately seven children.
E. Mali's population is growing by approximately 2.7 percent each year.

Be sure the *total fertility rate – 2008* and *population growth rate – 2008* folders are turned off and turn on the *infant mortality rates - 2008* folder. Angola's very high infant mortality rate is a product of years of civil conflict, poor health infrastructure, and tropical diseases such as malaria. In 2008, nearly one in five children born in Angola did not make it to the age of one. The 2008 figures were exacerbated by a cholera epidemic in the country, but the layer shows that there are similarly high figures in other Sub-Saharan Africa countries. Infant mortality is often a good indicator of overall development. Explore the world to get a sense of typical infant mortality rates in the rest of the world.

3. Identify the six countries with the highest infant mortality rates in 2008. What do these countries have in common?

A. They all have very high HIV/AIDS rates.
B. They are all in Sub-Saharan Africa.
C. They all have high incidences of malaria because of their tropical locations.
D. They all lost infants to the 2006 Indian Ocean tsunami.
E. They have all had internal or adjacent civil conflicts in the last 20 years.

Not surprisingly, there is a strong relationship between infant mortality rates and our next theme to explore, life expectancy. Turn off *infant mortality rates – 2008* folder and turn on *life expectancy at birth - 2008*. Life expectancy tells us the average number of years we would expect someone to live if they were born today and conditions did not change in the future. It's another good measure of development or overall quality of life in a country.

4. Regarding life expectancy in Sub-Saharan Africa, which of the following statements is *not* accurate?

A. At least one country's life expectancy is below 40.
B. The lowest South American country's life expectancy is higher than the highest Sub-Saharan African country (excluding island countries).
C. Life expectancy in Sub-Saharan Africa's most affluent country (South Africa) is more than 30 years less than in Japan.
D. Life expectancies are even lower in North Africa than Sub-Saharan Africa.
E. As a region the average life expectancy for Sub-Saharan Africa is less than 60.

Turn off the *life expectancy at birth – 2008* folder. One effect of Sub-Saharan Africa's rapidly growing population is that the cities of the region are growing rapidly as well. For example, the region's largest city, Lagos, had a population of less than 300,000 in 1950. Today, it has population of at least 15 million. Open the 360° city scene of the *Lagos Oshodi Market* and study the scene. How does this market differ from your version of the market place, like a shopping mall? This looks like a large market, but get a better feel for it by exiting the photo and zooming out to an eye height of approximately 800 meters. A major challenge that comes along with rapidly growing cities is providing adequate infrastructure. It's difficult for the road, sewer, water, and electrical services to keep pace. Open the 360° city scene of the Lagos internet café. Check out the electrical service pole in the foreground of the image. Do you think you would see something that looked like that in your neighborhood?

5. What generalization can you make after viewing the *Lagos Oshodi Market* scene?

A. There is a large demand for taxi/van service.
B. Western chain stores dominate the market's retail atmosphere.
C. Women in Lagos prefer to wear pants.
D. Nigerians prefer to wear clothing in muted earth tones.
E. Private automobiles are not allowed in the market area.

Exploration 6.2: SHORT ESSAY

1. Study the Lagos Oshodi Market scene. Compare this scene with a typical shopping trip for you to your market or mall. What cultural differences can you identify? How would an experience shopping at this market be different than an experience at your typical market or mall?

____
____
____

2. After examining the *Lagos internet café* city scene, identify at least three features that suggest that infrastructure is not as developed in Lagos as it is in the typical North American city. Beyond listing your observations, provide some insight as to the hazards or detrimental effects of these infrastructural shortcomings.

____
____
____

Exploration 6.3: SUB-SAHARAN AFRICA CULTURE

Verify that the sub-folders of Sub-Saharan Africa are turned off and then turn on and expand the 6.3 CULTURE folder. Sub-Saharan Africa has a rich cultural heritage that includes a wide array of religious practices. The region is home to a number of nature-based religious traditions that can be collectively described as Animism. As with any religious tradition, there are innumerable variants of Animist traditions. From a globalization perspective, the Animist traditions are noteworthy because they were brought to the Americas by West African slaves. These ideas and practices were the basis for the syncretic religions that would arise in the Caribbean and Plantation America. These religions, like Santería, meld the Yoruba practices of West African traditions with Roman Catholic and indigenous traditions. Go to the *Ouidah, Benin* placemark and utilize Google Earth™ and outside resources such as your textbook to determine the religious significance of this location in terms of syncretic religions. You will find it helpful to turn on the *Panoramio Photo* layer and explore the images in the vicinity.

Exploration 6.3: MULTIPLE CHOICE

1. Which of the following syncretic religions can trace its heritage to Ouidah, Benin?

A. Druze
B. Candomblé
C. Caodaism
D. Rastafari
E. Haitian Vodou

Christianity is the majority religion in many countries of the region, but only gained this prominent role in the wake of European colonialism. The Ethiopian Orthodox Christian tradition represents one of the few locations where the presence of the church predates colonialism. Lalibela is the second holiest city in Ethiopia and is a prominent destination for religious pilgrims. Zoom to the *Lalibela, Ethiopia* placemark and utilize imagery and any additional materials such as the *Panoramio Photo* layer to identify what is so unique about this location.

2. Lalibela, Ethiopia is home to some very unique Ethiopian Orthodox churches. They are unique because they

A. are very tall and narrow structures.
B. are built on the peaks of steep hills.
C. are constructed almost entirely of salt
D. are carved from solid rock.
E. are larger than almost every church in the world.

Turn on the *3D Buildings* layer, then click the *Basilica of Our Lady of Peace, Côte d'Ivoire* placemark. This remarkable structure is the largest church in the world. The building reportedly cost more than $300 million when it was completed in 1989.

3. The Basilica of Our Lady of Peace in Yamoussoukro, Côte d'Ivoire bears a striking resemblance to another famous church. Use Google Earth™ to identify which of the following churches is most similar.

A. the Basilica of St. Peter in Vatican City
B. St. Paul's Cathedral in London
C. Notre Dame Cathedral in Paris
D. Hagia Sophia in Istanbul
E. St. Basil the Blessed in Moscow

While the cost of the construction of the *Great Mosque of Djenné* is unknown as it was built in the 13th century, it too qualifies for a world's largest designation. Study the building, the site and the surrounding environment and identify what makes this structure unique.

4. The Great Mosque of Djenné is the world's largest

A. glass building.
B. floating building.
C. mud brick or adobe building.
D. concrete building.
E. steel building.

Finally, click the *National Mosque of Abuja, Nigeria* placemark. Islam in Nigeria is an interesting point of study because Nigeria lies in the cultural transition belt where Islam is prevalent to the north and Christianity is prevalent to the south. There have been religious undertones to strife in the country in recent years. The mosque can be a very distinctive feature on the landscape that helps establish a cultural presence. This mosque has distinctive architectural features that are commonly associated with Muslim places of worship.

5. The National Mosque of Abuja, Nigeria has <u>four</u> distinctive features associated with Islamic mosques. These are known as

A. flying buttresses.
B. minarets.
C. naves.
D. transepts.
E. domes.

Exploration 6.3: SHORT ESSAY

1. Some Animist traditions remain viable and heavily practiced in Sub-Saharan Africa. However, these animist traditions usually do not have elements of the built environment that are as conspicuous as those associated with the universal religious traditions of Christianity and Islam. For example, click the *Ife, Nigeria* placemark. Ife is significant to the Yoruba people because this is believed to be the location where the deities Oduduwa and Obatala commenced creation of the world. Utilizing the *Panoramio* photos and outside resources such as your text and the internet, how would you characterize the religious landscape of this focal point of the Yoruba's religious tradition in comparison to the Christian and Muslim sites highlighted in this exploration?

____
____
____

2. Animist, Christian, and Islamic religious traditions have been addressed in this exploration, but what about other major world religions? Can you identify a location within Africa that has adherents of Judaism and another that has adherents of Hinduism? What are the historical or contemporary explanations for the distribution of the followers of these faiths in Africa?

____
____
____

Exploration 6.4: SUB-SAHARAN AFRICA GEOPOLITICS

Inside Sudan is an area of subsistence farming and nomadic farming known as Darfur. This ethnically diverse region has been home to a conflict since 2003. Government-armed militia, known as janjaweed, have specifically targeted ethnic groups that had a history of supporting rebel groups in the area such as the Sudan Liberation Movement (SLM). Civilian casualties in the region have been immense, with more than 400 villages completely destroyed, millions of people forced to leave their homes, and several hundred thousand dead. Nearly three million remain displaced within the country of Sudan and another 300,000 refugees living in camps across the border in Chad. The United Nations estimates that nearly five of the six million persons living in Darfur continue to be impacted by this ongoing situation.

Expand the 6.4 GEOPOLITICS folder and turn on the *internally-displaced persons – 2006* folder. You can see that the number of internally displaced persons in the entire country of Sudan exceeded five million persons in 2006. Internally displaced persons (IDP) are distinguished from refugees in that they have not crossed an international border as a result of their displacement. By no means does this mean that they face less hardship than an official refugee. In fact, it can be harder for these persons as they are not afforded the same legal status and protection as a refugee. Refugees and IDP figures can vary dramatically from one year to the next as a result of evolving conflicts and governance. A great place to learn more about IDPs is the Internal Displacement Monitoring Centre (IDMC). The IDMS monitors IDPs continuously and provides statistics on their status. Identify the five African countries with the highest IDP figures and then open the link to the IDMC. From the IDMS homepage, view the IDP world map. Zoom to Africa and examine the most current figures for the five countries you have already identified.

Exploration 6.4: MULTIPLE CHOICE

1. What African country had one of the five highest IDP counts in 2006, but had a much lower figure by 2008 as evidenced by data from the IDMC?

A. Chad
B. Central African Republic
C. Democratic Republic of the Congo
D. Uganda
E. Zimbabwe

2. Based on the *internally displaced persons – 2006* folder and the data available from the IDMC which of the following world regions has the most internally displaced persons?

A. East Asia
B. Europe
C. Latin America
D. Southeast Asia
E. Sub-Saharan Africa

Turn off the *internally-displaced persons – 2006* folder and expand and turn on the Darfur sites folder. Read the linked quotations and view the linked photographs, and then turn on the IDP and refugee camps folder. Click the folder title to gain a perspective view. These bars represent the relative size of displaced persons encampment.

3. All of the tactics listed below have been utilized by the janjaweed in the Darfur conflict except

A. rape.
B. poisoning wells.
C. gouging out eyes.
D. burning villages.
E. fining members of certain ethnic groups.

4. Based upon what you have learned in this exploration, what is the difference between the purple bars (the more eastern ones) and the blue bars (the more western ones)?

A. Blue represents citizens of Chad while purple represents citizens of Sudan.
B. Blue represents citizens of Sudan while purple represents citizens of Chad.
C. Blue represents internally displaced persons while purple represents refugees.
D. Blue represents refugees while purple represents internally displaced persons.
E. Blue represents deaths while purple represents injuries.

Finally, turn off the 6.4 GEOPOLITICS folder and explore the landscapes of the Darfur region. Think about what challenges this would present for refugees and internally displaced persons.

5. The Darfur crisis is unfolding in an area that can be described as

A. dense forest.
B. swamps and marshes.
C. fertile grasslands.
D. high mountains.
E. semi-arid to arid.

Exploration 6.4: SHORT ESSAY

1. Research the position and actions of your country in reference to Darfur. Do you feel you country should be doing more or less? Why?

____
____
____

2. Big political changes may be coming to the Darfur region in 2011. Additionally, the leader of Sudan may be held responsible for the atrocities of Darfur by the International Criminal Court. Provide an update to the current situation in Darfur.

____
____
____

If you are interested in exploring this topic further, turn on the *Crisis in Darfur* folder. This folder contains the sites and links you have already viewed plus many more. This content was produced by the United States Holocaust Memorial Museum.

Exploration 6.5: SUB-SAHARAN AFRICA ECONOMY AND DEVELOPMENT

Open the 6.5 ECONOMY AND DEVELOPMENT folder and explore the *Cullinan* and *Kimberley* placemarks. These sites are famously tied to what has been a leading industry in South Africa for more than 150 years. Utilize Google Earth™ to determine what type of industrial activity takes place at these locations.

Exploration 6.5: MULTIPLE CHOICE

1. The Cullinan and Kimberley placemarks highlight two locations in South Africa that are known for

A. coal mining.
B. diamond production.
C. industrial waste disposal.
D. concrete production.
E. nuclear testing.

More recently, petroleum resources have caught the attention of oil companies and emerging national powers from around the world. Increasingly, this region's oil and gas resources are being developed. The West African country of Nigeria has seen some of the most significant development of these resources. Along with the development, however, have come significant challenges. Some of the Niger Delta's inhabitants have clashed with foreign oil companies over issues of exploitation and pollution. Go to the *Port Harcourt* placemark and then click the link to the National Geographic article on African oil development. Read the article and be sure to look at the associated features such as the photo gallery, map, and did you know? Now click the *gas flares* placemark and you will see one of the environmental impacts of oil development in the Niger Delta. These fires are burning off natural gas that is in this case an unwanted by-product of oil production. Open the link associated with *gas flares* placemark. The link takes you the US Energy Information Administration. If you ever need energy-related information, this is the place to find it.

2. Based upon information you have gathered from the resources linked to this exploration, identify the top oil-producing country in Africa in 2004 and in 2008.

A. Nigeria in 2004, Algeria in 2008
B. Sudan in 2004, Chad in 2008
C. Algeria in 2004, Angola in 2008
D. Chad in 2004, Sudan in 2008
E. Nigeria in 2004, Chad in 2008

Africa uses its natural resources for the production of numerous agricultural commodities. While traditional, low-technology, low-capital methods are employed in some locations in the region, there is also plenty of evidence to suggest that high-tech and capital-intensive methods are utilized as well. For example, go to the *sugar plantation* placemark and you will see a modern South-African agricultural operation. On the other hand, traditional methods of economic production can still be spotted on the landscape. Click the *ancient commodity* placemark and study both the high-resolution imagery and the surrounding landscape to deduce what you are viewing.

3. The *ancient commodity* placemark highlights the production of

A. shrimp via aquaculture.
B. cotton.
C. sea salt.
D. silk.
E. opiates.

Yet another ancient commodity can be viewed at the *primary sector export* placemark. Unfortunately, the harvesting of this commodity is not taking place at a sustainable rate in this part of Sub-Saharan Africa. Utilize the clues of the scene to identify the resource.

4. The *primary sector export* placemark displays the export of

A. coal
B. cigars
C. seaweed rolls
D. steel pipe
E. timber

Turn on the *age structure (% 0 to 14 years of age) – 2008* folder. It is apparent that Sub-Saharan Africa is the globe's youngest region as a significant percentage of the population falls within this category. This can present tremendous opportunities as there will soon be many more available workers, but it can also present challenges. First and foremost, there are very high numbers of dependents across the region. A high-dependecy ratio can be a real hindrance to the development of economies because too much of the adults energies are directed at providing for very young or very old populations. This issue is exacerbated when you consider the HIV-AIDS pandemic that disproportionately impacts young adults. Turn off the *age structure (% 0 to 14 years of age) – 2008* folder and turn on the *HIV/AIDS prevalence rate – 2008* folder. Compare the *age structure* and *HIV/AIDS* datasets by toggling the folders off and on. Think about the implications for the children where disability and mortality rates associated with HIV/AIDS are high.

5. What specific part of the Sub-Saharan Africa region appears to be most at-risk for high numbers of orphans due to HIV-AIDS?

A. central
B. eastern
C. northern
D. southern
E. western

Exploration 6.5: SHORT ANSWER

1. Research current economic trends in Sub-Saharan Africa (SSA). Briefly address the following inquiries: How is SSA's economic performance compared to other world regions since the global economic slowdown commenced in 2008? What are a few of the sectors/industries that have demonstrated the most growth in SSA in recent years? What are the long-term economic prospects for the region?

____

____

____

2. Address the challenges facing nations with high percentages of dependent youth with simultaneously high rates of HIV-AIDS infection. Discuss what countries have been most affected by this problem and what is being done in terms of policies or programs to ameliorate the situation.

____

____

____

YOU MAP IT! Africa's Wildlife

Africa is known for its diverse and large fauna. The lions, wildebeests, and giraffes are not only important ecologically, but also economically. Wildlife-based tourism injects millions of dollars into local economies on an annual basis. While most of the imagery utilized for Google Earth™ is too coarse to view individual examples, there are exceptions. For example, Michael Fay's *Africa Megaflyover* layer is contained in the *National Geographic Magazine* folder which is located in the *Gallery* folder of the *Primary Database*. This has hundreds of large-scale, high resolution photographs from around Africa. Utilize this resource and any of the other resources of Google Earth™ (e.g. Panoramio photos, Jane Goodall's Gombe Chimpanzee Blog, WWF Conservation Projects) to take us on a guided tour of a sample of Africa's wildlife. Incorporate at least 5 photos, scenes, or links on your tour and provide a short narrative for each stop. Look in the YOU MAP IT! folder for an example image. An example narrative follows.

Hippos in the Mud.

These hippopotami were photographed in the drying mud of a river in western Tanzania. The hippo is the second-largest land animal behind the elephant. Hippos stay cool during the day by laying in the water or mud. It is not surprising to see this picture of hippos in Tanzania because it is second only to Zambia in the estimated number of wild hippos, according to the IUCN.

Turning it in:

Your instructor will provide you with an explanation of how to submit your results from this assignment. This may include e-mailing a .kmz file you create from your *YOU MAP IT!* folder and a document with the narrative to your instructor.

To create a .kmz from your *YOU MAP IT!* folder, simply click once on the *YOU MAP IT!* folder to highlight it, then go to File, Save Place As…, and save it in an appropriate location on your computer.

Name: ______________________________

Date: ______________________________

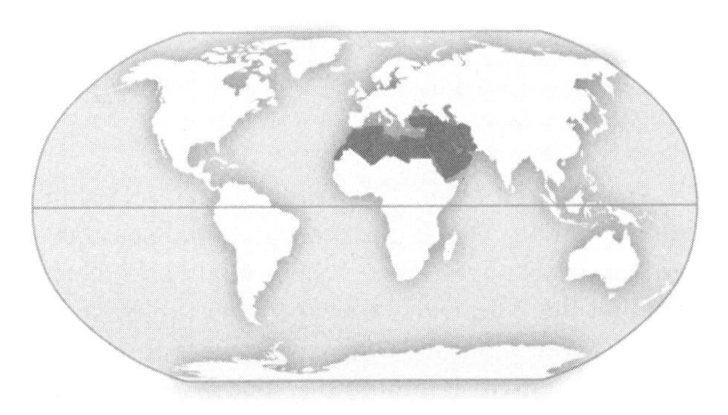

Chapter 7: Southwest Asia & North Africa

Southwest Asia and North Africa is a region most strongly characterized by arid and semi-arid climates, highly clustered populations, Islam as the dominant religion and Arabic as the dominant language, fossil-fuel production, and political instability. Physically, the region includes the part of Africa north of the Sahel, and the Arabian Peninsula. Its northern extent includes Anatolia and its easternmost state is the Islamic Republic of Iran.

The dry nature of the majority of the region is its defining environmental characteristic. Thus, the acquisition, allocation, and delivery of water are the primary concerns for many persons and governments around the region. As a result of this climatic reality, there have been some rather ambitious public works projects aimed at water development. Not surprisingly, the locations that are wetter and/or cooler are desirable locations for human settlement.

The population of the region is largely clustered around water resources such as coastlines, river valleys, and oases. Cities are densely populated and growing rapidly. This has resulted in some unique living arrangements, such as Cairo's City of the Dead. Tehran, Cairo, and Baghdad are the region's largest cities. Smaller clusters of people can exist in the harsh deserts of the region if they have access to a spring or aquifer.

This region is the hearth area of the large monotheistic traditions of Judaism, Christianity, and Islam. Whatever the root causes, this region is the point of origin for these religious traditions. Sites like Mecca, Medina, and Jerusalem are destinations for pilgrims from around the world. Jerusalem is a remarkable cultural landscape because so many important holy sites are located in close proximity.

This proximity helps explain some of the seemingly ongoing conflict inherent to this part of the globe. Recent years have produced the Iraq War, unending tensions between Palestinians and Israelis, and a growing uneasiness about the militarization of nuclear technology in the region.

The rich energy resources in the region are yet another source of conflict as states within and outside the region have sought to control the vast deposits. On the other hand, these resources have brought great prosperity to some individuals and government regimes in the region. This wealth is on display in extreme ways in the United Arab Emirates state of Dubai. One such way is the proliferation of unique skyscrapers reaching into the desert sky.

Download EncounterWRG_ch07_SouthwestAsiaNorthAfrica.kmz from ***www.mygeoscienceplace.com*** *and open in Google Earth™.*

Exploration 7.1: SOUTHWEST ASIA & NORTH AFRICA ENVIRONMENT

While oil may be the first resource that comes to mind when one thinks of Southwest Asia and North Africa, water is a more crucial resource on a day-to-day basis for the inhabitants of the region. The region's historical, contemporary, and future trajectories in the realms of population, geopolitics, and economic development are tied to the distribution and control of this resource.

Any discussion of water in the region should reference Egypt's *Aswan High Dam*. This dam was completed in 1970 and has profoundly impacted the livelihoods of Egyptians. The dam was built to regulate the seasonal flooding of the Nile, provide water for agricultural development, and electricity. While these goals have been achieved, the construction of the dam means that the floodplain and the Nile Delta are no longer replenished with nutrients and silt on an annual basis. As Lake Nasser filled, many thousands of people were forced to move and innumerable cultural treasures were consigned to a watery grave. Examine the *Aswan High Dam* placemark and then explore the Nile River Valley, both upstream and downstream from the dam.

Exploration 7.1: MULTIPLE CHOICE

1. The Aswan High Dam backs water up behind the dam for a considerable distance in Lake Nasser/Lake Nubia. Use the ruler to measure the distance upstream that the river channel has standing water. Be sure to follow the channel when measuring.

A. <100 km
B. 100-200 km
C. 200-300 km
D. 300-400 km
E. >400 km

2. Assess the Aswan High Dam and the Nile River Valley. Which of the following statements is *not* accurate?

A. Irrigated agriculture is much more common upstream from the Aswan High Dam than downstream.
B. There is a second smaller dam downstream from the Aswan High Dam.
C. Canals are prevalent between the Aswan High Dam and the Nile Delta.
D. There are more people living along the river downstream from the dam than upstream.
E. Intensive agriculture takes place on the Nile Delta.

A more recent and even more ambitious project is the Libyan government's Great Man-made River Project (GMMR). The GMMR exploits the Nubian Sandstone Aquifer System, a fossil aquifer. The water was stored in this aquifer when the region was much moister. Radiocarbon dating indicates that the water in the aquifer is more than 40,000 years old in some cases. The *GMMR well-field* placemark illustrates some of the wells that have tapped this resource. This project is bringing this ancient water to the cities of Libya and to newly emerging agricultural areas. Expand the *Great Man-made River* folder and identify changes that have occurred to the respective landscapes in the 30 to 40 years by examining the three image pairs associated with the project. Double-click an image to zoom to the respective site and then toggle the images on and off. Pay attention to facets such as urban expansion, the growth in areas under cultivation, and construction of infrastructure such as roads.

3. The most significant infrastructure change in the Great Man-made River image pairs is

   A. the construction of an airport in the Tripoli imagery.
   B. the construction of an eight-lane super highway in the Suluq imagery.
   C. the construction of reservoirs in the Suluq imagery.
   D. the construction of an airport in the Al Kufra imagery.
   E. the construction of reservoirs in the Tripoli imagery.

4. With respect to agricultural change in the Great Man-made River image pairs, which of the following statements is *not* accurate?

   A. Suluq has expanded center-pivot and canal irrigation.
   B. Al-Kufra is still waiting for a road connection to nearby towns.
   C. Al-Kufra had experienced the most dramatic expansion in center-pivot irrigation of the three sites.
   D. The urban footprint of Tripoli has roughly doubled.
   E. Agriculture is being practiced on former dunes southeast of Tripoli.

As geographers, we often try to develop better understandings of the complex world by creating regions. It's important to remember that any generalization can obscure important real-world realities. For example, a central theme of Southwest Asia and North Africa is the arid nature of the region. It's easy to think of the region looking exclusively like Saudi Arabia's *Empty Quarter*. However, much of the region is not a flat, dry desert. Double-click the *atypical landscapes* tour. The tour may begin to play before it has finished loading. Repeat the tour after it has completed playing to see it at a higher-resolution.

5. Watch the atypical landscapes tour and identify which of the following statements is not accurate about this part of the Southwest Asia and North Africa region.

   A. There are rain-fed agricultural fields.
   B. There are seaside beach resorts.
   C. There are dense mountain forests.
   D. There is no evidence of frozen precipitation.
   E. There is a mountain peak that exceeds 5,500 meters.

Exploration 7.1: SHORT ESSAY

1. The human imprint in and along the Nile River Valley is a distinctive element of the world's cultural landscape. It can be discerned from several thousand kilometers above the surface of our planet. Study the Southwest Asia and North Africa region and identify the most analogous feature. What makes your comparative feature similar and how is it different?

   ____
   ____
   ____

2. The Great Man-made River (GMMR) project has been controversial for a number of reasons including the fact that fossil water that will not be replenished in many human lifetimes is being utilized. What do you think of the GMMR? Is it a wise use of resources to promote development? Provide at least three reasons why you support or oppose the project and at least one reason why the converse position may be valid.

____

____

____

Exploration 7. 2: SOUTHWEST ASIA & NORTH AFRICA POPULATION

Double-click and then expand the 7.2 POPULATION folder. Turn on the *population density* layer. It will take a few seconds to load. The scale is in the Indian Ocean. One can see that the population in the region is distributed very unevenly. For example, the most populous country in the region has a population that is clustered around the Nile River and its delta. The majority of the country has very low population density. In a region where water is such an important variable, people cluster around rivers, coastlines, and oases. Study the *population density* of the region and identify which countries have the most clustered and the most evenly distributed populations.

Exploration 7.2: MULTIPLE CHOICE

1. Identify the statement that is most accurate in terms of population density.

A. North African countries are densely populated in their interiors and sparsely populated on the coasts.
B. The Persian Gulf coastal area is the most densely populated part of Iraq.
C. Most of Libya has lower population density than the West Bank.
D. Western Yemen is the least densely populated portion of the Arabian Peninsula coast.
E. The population of Turkey is more evenly dispersed than that of Saudi Arabia.

Turn off the *population density* layer. Urbanization in this region has accelerated markedly in the last 20 years. In 1990, 44 percent of the region's population lived in urban settings. Today that number has risen to 59.9 percent. By 2020, it is anticipated that 70 percent of the population will be city dwellers. As in other parts of the world, this rapid urbanization presents unique infrastructural challenges. Often times, however, the city is expanding into prime agricultural lands or forested areas. Open the *Damascus, Syria* folder and compare the images from 1972 and 2005. You can see that the gray urban area has expanded at the expense of the green agricultural areas. A similar phenomenon has occurred in and around Cairo, Egypt. The *Cairo, Egypt* folder contains a series of three images that illustrate the urban expansion of a city that has expanded its population from one million in 1930 to more than 16 million today. When you finish examining the urban changes in Cairo, turn off the *Cairo, Egypt* folder. This increasing population has led to some unique living circumstances for some residents of Cairo. Click the *City of the Dead* placemark and study the landscape. Consult outside resources to determine what you are seeing. *Panoramio* photos will provide some additional insight into the landscape as well.

2. In the period from 1972 to 2005, describe the expansion of Cairo's urban footprint.

A. Predominantly north into desert areas.
B. Spatial footprint has not increased, but density has.
C. Retreated from agricultural areas and grown in arid areas.
D. East into the desert and north into the delta.
E. Predominantly west into arid areas.

3. At least partially due to increased population pressures, Cairo's City of the Dead

A. is a new high-density condominium development.
B. is an area of condemned buildings that are being torn down to build palatial estates.
C. is a giant cemetery where thousands of people live among the dead.
D. is the primary commercial district.
E. is the largest known commune Sufi Muslims.

One can also identify highly clustered populations, albeit at much smaller scales across the deserts of the region. In these cases, a water resource is available, often in the form of a spring. Vegetation and people are thus found in places that you hardly expect them. There are three oasis-related placemarks that provide you with a sample of how a Saharan Oasis can appear. Examine these locations, noting similarities and dissimilarities between the sites. The *Panoramio Photos* can be useful here too.

4. Examine the three oasis-related placemarks and select the statement that is *not* accurate.

A. Tozeur does not have the surface water that is evident at Behariya.
B. Mandara Oasis has the greatest area dedicated to agriculture.
C. Tozeur has the largest urban imprint of the three oases.
D. Bahariya oasis is associated with the greatest local variation in relief.
E. Mandara has the smallest urban footprint of the three oases.

The move to more urban-based settlements often goes hand-in-hand with increasing levels of development. Perhaps the most well-known measure of human development is the United Nations Human Development Index (HDI). The HDI is a calculation based on life expectancy, knowledge and education levels, and standard of living. Open the *Human Development Indices* link. The Human Development Index Trends Motion Chart allows you to manipulate relevant variables to illustrate human development from regional and national perspectives. Utilize the motion chart with its preset variables and watch the animation by clicking the play button. Now change the "Lin" drop-down box to "Log" and watch it again. If you click a region it will make those countries flash during the animation. Assess the change in human development over the period illustrated.

5. Select the generalization about development in the Southwest Asia/North Africa region (aka Arab States) compared to other world regions from the period 1980-2007 that is *not* accurate.

A. Arab states have generally not improved as much as Sub-Saharan African states.
B. As of 2007, Kuwait has the highest development levels of the Arab states.
C. As of 2007, Djibouti has the lowest HDI of the Arab states.
D. Arab states have seen more numerical improvement on the HDI scale than OECD states.
E. Egypt HDI has progressed more than Syria's over the study period.

Exploration 7.2: SHORT ESSAY

1. The Gaza Strip and Bahrain are the two most densely populate entities in the region. Use Google Earth™ to zoom-in to these respective locations. Compare and contrast the two locations. How is the land divided and used differently? What looks like a more affluent population? In what condition are the major infrastructural components?

    ____
    ____
    ____

2. Use the Human Development Index Trends Motion Chart to uncover trends in development between two countries in the Arab States region from 1980-2007. Utilize drop-down boxes along the x-axis and y-axis to change the variables. Describe the trends in least three variables for the countries you selected. Hypothesize as to what might explain the patterns you see.

    ____
    ____
    ____

Exploration 7.3: SOUTHWEST ASIA & NORTH AFRICA CULTURE

Perhaps no religion is more strongly associated with a region than Islam is with Southwest Asia and North Africa. Islam is evident in multiple facets of the cultural landscapes of the region. From language to places of worship to attire, the dominant religion reveals itself. One cultural facet that speaks to this phenomenon is the prevalence of the sacred color of Islam. Open the *flags* link and peruse the flags of the world. Observe the flags of the Southwest Asia and North Africa region. You should notice one color that is more common than any other. Another good place to look for Islam's sacred color is inside a mosque. If you fly into the *Alasahan* Mosque in Bagdad, Iraq you can verify your assessment.

Exploration 7.3: MULTIPLE CHOICE

1. The most common color associated with Islam is

    A. blue.
    B. green.
    C. orange.
    D. purple.
    E. yellow.

While Mecca and Medina may come to mind first when one thinks of Islam's holy sites, Jerusalem is also very important. The Dome of the Rock commemorates the spot where according to Islamic tradition, Muhammad ascended to heaven with the angel Gabriel. Click the *Dome of the Rock* placemark to see this sacred site. Turn on the *3D Buildings* layer to gain a better perspective on the structure. Now fly into 360° scene of the *West Wall of Jerusalem.* Manipulate the image so you can see the large stone wall. This wall is the west wall of the Temple Mount. Followers of Judaism come to this very holy location to pray. This wall is a remnant of the Second Temple that occupied this site. It is the Western Wall of what is referred to as the Temple Mount. Now exit the photo and back out to the point that you can see the Dome of the Rock. Use the ruler tool to measure the distance between the Western Wall and the Dome of the Rock.

2. The distance between the Western Wall and the Dome of the Rock is approximately

   A. 150 meters.
   B. 1.5 kilometers.
   C. 15 kilometers.
   D. 150 kilometers.
   E. 150 kilometers.

Now that you have a feel for the distance between two of the most important locations for two of the world's Abrahamic religions, let's think about Christianity. Do Christians have important sites in and around Jerusalem? Now fly into 360° scene *Church of the Holy Sepulchre*. Many Christians believe this is the location of Golgotha, the site where the New Testament says Jesus was crucified. There are many other holy sites such as the *Chapel of the Ascension*. This is a holy site for both Christians and Muslims where one can see the alleged footprint of Jesus. This vicinity has an unmatched density of noteworthy religious sites. While this can be rewarding for the religious tourist, it presents great challenges for governance in the region. A geopolitical reality of this location complicates matters even more.

3. Utilize Google Earth™ to evaluate which of the following statements is accurate.

   A. The Gaza Strip is adjacent to Jerusalem.
   B. Armenia, Israel, and Syria claim Jerusalem.
   C. The Jordan River, dividing Israel and Jordan, runs through Jerusalem.
   D. Jerusalem is located on the border between Israel and the West Bank.
   E. Jerusalem is part of the Golon Heights, which is claimed by Syria.

Religion in the region also strengthens patterns of language. Judaism and Islam are both strongly associated with respective languages. Hebrew is the language of the Torah, the Hebrew Bible, and is an integral part of Judaism. Fly into the *Sana'a Yemen 1* and *Sana'a Yemen 2* placemarks and you will see several examples of the script of the language that is tied to Islam.

4. The language of Islam is

   A. Muslim.
   B. Cyrillic.
   C. Mosque.
   D. Yemen.
   E. Arabic.

Religion and language aren't the only cultural differences between this region and the Americas. Basic cultural preferences can also be identified in something as mundane as sport. Double-click the *Golf Search* placemark and you will fly to Tehran, Iran. Tehran has a population of approximately 9 million people in the core of the city. How many golf courses are needed to fill the demand for millions of Iranians? For comparison's sake, Houston, Texas has more than 30 golf courses in a city with an urban population less than half of Tehran's. Zoom in to the area enclosed in the brown polygon and study the landscape.

5. How many golf courses are found within the Golf Search polygon of Tehran?

   A. one course with less than 18 holes
   B. one 18-hole course
   C. five 18-hole courses
   D. 15 to 20 18-hole courses
   E. more than 30 18-hole courses

Exploration 7.3: SHORT ESSAY

1. Return to the Sana'a Yemen scenes. Study the scenes carefully and list at least three cultural elements that are different than what you see on a daily basis. Include a discussion of the daggers as one of your elements. Also illuminate and discuss one cultural element that is not different or foreign to you.

____

____

____

2. Revisit the Old City of Jerusalem and think about the complex religious and geopolitical situation. Do you feel there are more opportunities for unity and common ground or division among the Abrahamic traditions as a result of this geography? Explain your response.

____

____

____

Exploration 7.4: SOUTHWEST ASIA & NORTH AFRICA GEOPOLITICS

Southwest Asia and North Africa present a complex and dynamic palate for the study of geopolitics. The past, current, and future conflicts in the region can be seen clearly through the lens of Google Earth™. The largest-scale conflict of the past decade has been the US-led war in Iraq. Turn on the historical imagery tool by clicking the clock icon in the toolbar, and then click the *Green Zone* placemark. The term Green Zone refers to the International Zone in Baghdad, Iraq that has been home to a large proportion of the international entities operating in post-invasion Iraq. It was one of two primary garrisons for occupation forces. Study imagery from 2002 and then utilize the time slider to advance the imagery to September 6, 2004.

Exploration 7.4: MULTIPLE CHOICE

1. What phrase could explain the difference in the structure where the Green Zone placemark and label are located in this imagery?

   A. expanded office space
   B. aerial bombardment
   C. cleared for parking
   D. truck bomb
   E. conversion to mosque

Zooming to the Baghdad Airport placemark will return you to March 31, 2002 and an intact hub of transportation infrastructure. Use the time slider to move forward to September 25, 2003. Zoom in and out as necessary to survey the airport, and identify what has changed from one image to the other.

2. From March 31, 2002 to September 25, 2003 the Baghdad Airport

A. became a deployment area for at least 35 helicopters.
B. had its runways completely destroyed.
C. increased the number of commercial airliners positioned at boarding gates by a factor of five.
D. had its main terminal expanded to more than twice its original size.
E. had two additional runways built.

Now look at the area that is adjacent to the northeast side of the airport. Use the time-slider to observe the differences between 2002 and 2005, and then turn off the historical imagery.

3. Which of the following statements is *most* accurate regarding the changing landscape northeast of the Baghdad Airport?

A. The area has been bombed to the point where most buildings have been destroyed.
B. It has been converted to a massive garrison.
C. It has been added to the airport because of the new extended runways.
D. The area has been converted to agricultural fields.
E. A moat has been added around the perimeter of the area.

The contemporary conflict in Iraq, albeit a major conflict by many measures, did not threaten to impact the flow of goods on a regional or global basis. This has not always been the case in the region. Although the duration of the conflict and the number of casualties pales in comparison to the Iraq War, a fascinating episode in geopolitical history involving the Suez Canal occurred in 1956. The nationalization of the canal by Egypt and a subsequent conflict involving Egypt, Israel, Great Britain and France temporarily stopped the flow of goods through the canal. Click the *Suez Canal* placemark to view the route in Google Earth™. Open the link to go to the Suez Canal Authority website. Utilize the information on the page and/or the "Saving in Distance" animations to evaluate how effective the Suez Canal is as a shortcut.

4. The Suez Canal provides the greatest distance savings for a ship traveling from

A. Singapore to New York.
B. Colombo to New York.
C. Ras Tanura to Rotterdam.
D. Tokyo to Rotterdam.
E. Jeddah to Piraeus.

Click the *Operation Orchard* placemark. On September 6, 2007 this site was completely destroyed by Israeli fighter-bombers. The US has stated that this was a nuclear facility with a military purpose. Look in the bottom, left-hand corner of your screen to see the date associated with this image. This was not the first time that Israel has acted to eliminate a potential nuclear threat in the region. Click the *Operation Opera* placemark. Operation Opera was a 1981 Israeli attack on a nuclear reactor being constructed at Osirak, Iraq. Some believe that the nuclear program of Iran could be the next target. Attacking Iran's nuclear facilities would be more challenging for Israel because the activities are allegedly distributed across multiple sites and Iran is a much farther target for Israel than Syria or Iraq. The location that is believed to be the center of Iran's enrichment program is near Natantz. Click the *Natantz Enrichment Facility* placemark to examine the location. Officials from Iran have stated publicly that they would shut off access to a key global choke-point for the flow of oil if the US and/or Israel were to attack their nuclear facilities.

5. Utilizing Google Earth™ and any additional outside resources, what is the global choke-point that Iran claims the ability to shut off if they are attacked in the future?

A. Panama Canal
B. Suez Canal
C. Bab-el-Mandab
D. Persian Strait
E. Strait of Hormuz

Exploration 7.4: SHORT ESSAY

1. Detail the changes that you can see in the Green Zone of Baghdad and in the area immediately northeast of the Baghdad airport. Identify the addition or deletion of features on the landscape and how this reflects the changing conditions of the Iraq War.

____

____

____

2. Briefly compare and contrast the three sites of alleged nuclear development included in this exploration. For example, do some sites appear to be going for covert development while others appear to emphasize fortifications and security?

____

____

____

Exploration 7.5: SOUTHWEST ASIA & NORTH AFRICA ECONOMY AND DEVELOPMENT

Open the ECONOMY AND DEVELOPMENT folder and click the *regional industrial landscape 1* placemark. Explore the landscape around this placemark. You will see that this is an area of hydrocarbon development in Kuwait. These oil fields look quite different than oil field developments in the United States. The numerous dark spots on the ground are actually standing oil. For environmental reasons this would be unacceptable in many of the world's more developed countries. Click the *regional industrial landscape 2* placemark and you will slide down the coast to another typical industrial site for this region. Zoom in and study the features of this site.

Exploration 7.5: MULTIPLE CHOICE

1. The most likely description of this industrial site is

A. solar power plant.
B. container ship transfer point.
C. oil port.
D. seashell packing plant.
E. cruise ship port.

Water resources, as discussed above, are a precious commodity in the region. The lack of freshwater presents challenges to agricultural and industrial development. Some countries with adequate financial resources have opted to tackle this problem in a unique way. Examine the facilities at the *Jeddah* and *Jubail* placemarks and determine what these facilities are. You may need to utilize outside resources.

2. Select the word or phrase that describes the activities at Jeddah and Jubail.

A. fusion
B. desalinization
C. gasification
D. oil conversion
E. nuclear

Some development experts believe this region needs better transportation connectivity with the rest of the world. Click the *Strait of Gibraltar* placemark. This is the point where the African and European landmasses come within 15 kilometers of one another. As you can see, there is currently no bridge connecting the two continents. There is also no tunnel to date because of the expense of such a project and the challenges posed by differing transportation systems. For example, the rail gauges are not the same in Africa and on the Iberian Peninsula. However, the idea of an undersea tunnel is not unprecedented.

3. Use Google Earth™ to evaluate which of the following cities is home to the gateway for an undersea tunnel.

A. Kerch, Ukraine
B. Skagen, Denmark
C. Barrow, Alaska
D. Kyle of Lochalsh, UK
E. Folkestone, Kent

One way this region has become more connected to the world in recent decades is through increased foreign direct investment. A leading country in this realm is the collection of emirates known as the United Arab Emirates (UAE). UAE has one of the most developed economies in the region and had one of the fastest growing economies in the world in the late 20th and early 21st century. To get a taste of this explosive growth, view the 360° *Abu Dhabi rooftop* scene. Abu Dhabi is the capital and second-largest city. How many skyscrapers appear to be under construction in this one image? Speaking of skyscrapers, turn on the 3D Buildings layer, exit the *Abu Dhabi rooftop* photo and then click on the *desert skyscrapers* placemark.

Dubai is home to extreme wealth and ostentatious displays of that wealth. This is reflected in the city's numerous and architecturally innovative skyscrapers. It seems as if each new entry to the skyline is an attempt to be more flamboyant than the last. Navigate through the 3D skyscrapers of Dubai utilizing your Google Compass and Hand Compass tools. Click on the 3D models of the buildings to identify them by name.

4. The tallest skyscraper in Dubai is

A. Almas Tower.
B. Princess Tower.
C. Burj al Arab.
D. the Index.
E. Burj Khalifa.

Skyscrapers aren't the only outrageous thing in Dubai. Open the *Ski Dubai* link and fly into the image. This is quite an impressive manufactured landscape. But to see even larger manufactured landscapes click the *Dubai* placemark. You see three large clusters of islands/peninsulas. Zoom in and study the features.

5. Which of the following statements regarding the island/peninsula features visible near the Dubai placemark is *not* accurate?

A. These are anthropogenic features.
B. The World has the lowest levels of development of the three island/peninsula clusters.
C. These are areas designed to house low-income citizens of the UAE.
D. These features alter the natural patterns of water movement along the coast.
E. Exclusive homes are located on the Palm Jumeira.

Exploration 7.5: SHORT ANSWER

1. Do you believe the tremendous oil resources of the Southwest Asia and North Africa region are a blessing or a curse? Provide concrete details and examples that back up your position.

____

____

____

2. Research The World development at www.theworld.ae. Be sure to read "The World Principles." Write a paragraph discussing the ways this project does or does not align with your personal value system.

____

____

____

YOU MAP IT! Skyscrapers

You have seen the impressive skyscrapers of Dubai. Now continue your tour of the world's tallest structures and create a map of their locations. The first thing you will need to do is find a list of the world's tallest buildings and structures. Be sure you identify what exactly you are mapping. For example, are the world's tallest structures in any category, the world's tallest skyscrapers, or a collection of world's tallest structures by category (e.g., skyscraper, minaret, dam, clock tower, etc.). Your map will include 10 world's tallest features.

Create a .kmz with a placemark for each feature that includes the name of the feature, its height, and the year it was built. Create a document with the same basic information in a tabular form. You can see two examples in the YOU MAP IT! folder and below.

10 of the World's Tallest Structures:
1). CN Tower - Tallest Concrete Tower; Toronto, Canada; 553 meters; 1976
2). GRES-2 Power Station – Tallest Chimney; Ekibastusz, Kazkhstan; 420 meters; 1987

Turning it in:

Your instructor will provide you with an explanation of how to submit your results from this assignment. This may include e-mailing a .kmz file you create from your *You Map It!* folder and a document with the narrative to your instructor. To create a .kmz from your *You Map It!* folder, simply click once on the You Map It! folder to highlight it, then go to File, Save Place As..., and save it in an appropriate location on your computer.

Name: ______________________________

Date: ______________________________

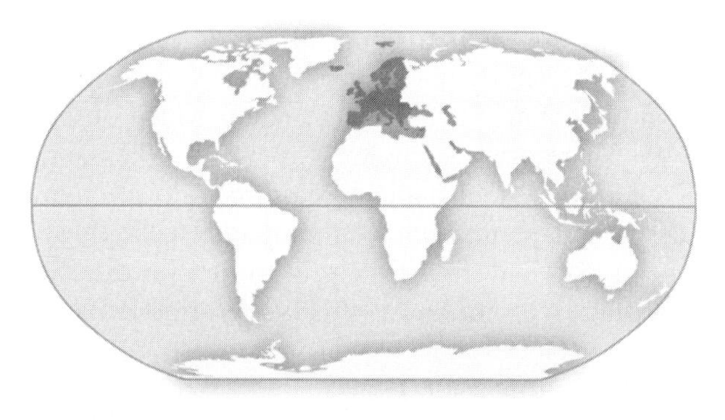

Chapter 8: Europe

Europe is the westernmost portion of the Eurasian landmass. The eastern edge of the region is often delimited by the Ural Mountains, Caspian Sea, and Black Sea. The Arctic Ocean lies to the north, the Atlantic Ocean to the west and the Mediterranean Sea to the south. Europe is home to approximately 50 countries and 750 million people. London and Paris are the largest cities that are undeniably European. Istanbul and Moscow, on the margins of the region, are larger.

Europe's environment is highly variable in terms of landforms and relief and Europe is home to some of the world's more ambitious water-management schemes. The Netherlands has extensively altered its physical landscapes to deal with the low elevations of the country while Venice, in Italy, also faces challenges related to sea level. Beyond the technical know-how and capital needed to bring public work projects to fruition, Europeans have been global leaders in developing an environmentally oriented ethic. A project located on European soil may provide agricultural salvation for the globe someday.

The population of the continent is aging with some countries already experiencing population declines. And yet Europe is one of the world's most densely populated regions. A fun way of examining population density in conjunction with levels of development is to view the night lights of settlements.

One of many settlements that can be identified via night lights is Paris, France- home to some of Europe's most recognizable cultural landmarks. Perhaps you have a mental image of the built environment of Paris. Google Earth™ can be utilized to verify and/or dispel and/or modify these mental maps. Paris is an interesting location to view a European perspective on cultural preservation. On the topic of culture in France, it would be hard to ignore the Tour de France. Le tour is one of the three most watched global sporting events along with the World Cup and the Olympics.

The Olympics have been held in Europe numerous times, including the 1984 games. Those took place in the Yugoslavian city of Sarajevo. Since the time of these games Sarajevo was subjected to an extended period of horrific warfare known as the Siege of Sarajevo. The country of Yugoslavia can't be found on the map anymore. In its place one sees a continually evolving geopolitical scene.

Our last exploration for Europe centers on a few of the technological contributions by citizens and states of the region. Europe has been and continues to be a global center of innovation, from developments in individual and mass transportation, to ongoing exploration of the final frontier.

Download EncounterWRG_ch08_Europe.kmz from **www.mygeoscienceplace.com** *and open in Google Earth™.*

Exploration 8.1: EUROPE ENVIRONMENT

While Europe is only one-third the size of Africa, its coastline is three times the length of the continent to its south. Coastal areas have always been magnets for populations because of their opportunities for economic gain. Therefore Europeans have a long history of interacting with the environment at the land-sea interface. Two locations that stand out in this regard are Venice, Italy and the country of the Netherlands. Much of the country of the Netherlands resides at or even below sea level. Therefore a complex system of dykes and pumps has been developed over the centuries by the Dutch. Kinderdijk is a particularly significant location in this regard. Zoom to the *Kinderdijk* placemark and explore the landscape to identify a particular technology that has been utilized in the area to help manipulate water levels. You may find it helpful to turn on the *3D Buildings* layer. You can also research the site via the World Heritage Committee, as this is a World Heritage Site.

Exploration 8.1: MULTIPLE CHOICE

1. At Kinderdijk, one can see the hydraulic technology of

A. wave-powered electrical generation.
B. the world's longest dam.
C. preserved windmills.
D. a network of steel sea walls.
E. a desalinization plant.

While the features at Kinderdijk serve the water management needs of the local area, the network of hydraulic management features in the Netherlands is extensive. Construction of these features began as early as the 9th century, but the big changes have occurred in the 20th and 21st centuries. The largest of the projects is illustrated in the two *Zuiderzee/Ijsselmeer* layers. Toggle these images on and off to illuminate what changed between 1973 and 2004. Be sure that *Borders and Labels* are turned on in the *Primary Database*. You can research the project further using outside resources, if necessary.

2. Between the 1973 and 2004 *Zuiderzee/Ijsselmeer*, all of the following changes can be noted except

A. a 20 to 30 percent increase in arable land.
B. construction of a new city (Almere).
C. Lake Ijssel has decreased in size by 50 to 60 percent.
D. construction of a lake in central Almere.
E. construction of a bridge east of Huizen.

Check to verify that the *3D Buildings* layer is on and zoom to the *Venice* placemark. The city of Venice stretches across more than 100 small islands in the Venetian Lagoon along the coast of northeastern Italy. The buildings of the city were built hundreds of years ago on wood pilings that were driven down into the clay of the seafloor. Extraction of groundwater for industry has caused the floor of the lagoon to slowly subside over the last century. Thus the buildings of Venice have become increasingly vulnerable to high tides. A number of solutions have been proposed, including refilling the aquifer, to building barriers across the entrances to the lagoon that can be closed when a particularly high tide is imminent. This latter solution is known as the MOSE project. In the Google Earth™ imagery, you can see modifications to the entrances to the Venetian Lagoon. Adjust your view as necessary and study the inlets to the Venetian Lagoon.

3. To complete the MOSE Project, the Italian government is building barriers to block the entrances to the Venetian Lagoon. How many lagoon entrances have to be blocked?

A. 2
B. 3
C. 4
D. 5
E. 6

While Europe has a reputation for its "green" perspective, the aforementioned projects have not been viewed favorably by many environmentalists because of the massive alterations they have brought to local ecosystems. Perhaps a better measure of Europe's environmental progressiveness can be seen through attempts to address environmental problems via policy. Not unlike significant portions of eastern North America, Europe has had an ongoing problem with acid rain. This acid rain has led to unacceptable levels of acidification in European environments. Open the link associated with the *critical acidification* placemark. This webpage has a variety of information related to the spatial distribution of problematic environmental situations in Europe. We are interested in figures five, six, and seven. These illustrate the change in critical acidification rates as a result of legislation that has been implemented in Europe. Study these maps carefully. You can click on each individual map to open a page with more information and the ability to magnify each map (click hard copy).

4. With respect to acidification in Europe, which of the following city pairs are projected to be most susceptible to problems in the year 2020?

A. Amsterdam Netherlands and Manchester UK
B. Manchester UK and Krakow Poland
C. Manchester UK and Rome Italy
D. Rome Italy and Krakow Poland
E. Amsterdam Netherlands and Krakow Poland

Now we will visit a very unique and quite important example of environmental foresight that is more tangible than the previously mentioned legislative changes. Go to the *environmental foresight* placemark. Utilize the tools of Google Earth™ to determine the significance of this sight.

5. Which of the following words or phrases is most directly related to the goals of the site associated with the environmental foresight placemark?

A. genebank
B. hydrology
C. cryonics
D. atmospheric ice cores
E. acidification

Exploration 8.1: SHORT ESSAY

1. Research the history of the transformation of the Zuiderzee into Lake Ijssel. Be sure to address the following questions in your response. What were the primary objectives of this project? What are the new lands called? What is the Markermeer?

___

___

___

2. Explain why it is a good idea to have a facility like the one associated with the environmental foresight placemark. Provide at least three reasons in your response.

 ____
 ____
 ____

Exploration 8.2: EUROPE POPULATION

Open the link attached to the POPULATION folder and learn about viewing cities at night from the scientists at NASA. After you have read the information from NASA, expand the *Gallery* folder in the *Primary Database*. Locate and expand the *NASA* folder, and then check the box next to *Earth City Lights*. This is a large data set, so it takes a little while to load. Study the Europe region and the world, at large. Identify patterns in the lights. Where are the greatest clusters? Where is it darkest? Is this purely a reflection of population, or do levels of development help explain the patterns, as well?

Exploration 8.2: MULTIPLE CHOICE

1. Based on your analysis of Europe at night, which of the following countries is most likely to contain a primate city?

 A. Spain
 B. Poland
 C. Italy
 D. Germany
 E. France

2. As you study the lightscape of Europe, you can identify areas of particularly high concentrations of light as well as areas of low levels of light production. Identify the following statement that is *least* accurate.

 A. The Po River Valley is one of the highest light-concentration zones.
 B. Albania is one of the most brightly lit countries.
 C. The coastal area of the Netherlands and Belgium is one of the highest light-concentration zones.
 D. Budapest appears to be a primate city.
 E. The Alps have clearly defined corridors of light passing through them.

3. Examine the North Sea. Explain the lights that are scattered across this body of water.

 A. These are cruise ships.
 B. These are schools of angler fish.
 C. These are islands.
 D. These are oil and gas platforms.
 E. These are boats harvesting Dungeness crabs.

Turn off the *NASA* folder and returning to the POPULATION folder, turn on the *median age – 2008* folder. Utilizing this data layer and the *population 65 and older (%) – 2008* folder, evaluate the demographics of the region with respect to old age. These layers clearly illustrate how Europe stands out with respect to older populations. The vast majority of the world's "old" countries are in Europe. When you have completed this exploration, be sure to uncheck the POPULATION folder.

4. After examining the two age-related European layers, what country outside the region is most similar in its proportions and median age of older citizens?

   A. Australia
   B. Brazil
   C. China
   D. Japan
   E. United States

5. In general, what European sub-region is younger?

   A. central
   B. eastern
   C. northern
   D. southern
   E. western

Exploration 8.2: SHORT ESSAY

1. Compare the night lights of Central and Western Europe to the night lights of the eastern United States. Identify differences between the two. Be sure to evaluate the spacing of towns and cities. Is there more regular spacing at one location or the other? How can you explain the patterns that you see?

   ____
   ____
   ____

2. Think of the ways that the older populations of Europe affect the decisions that European governments make. In particular, consider the ramifications of having many people that are retiring with many fewer young people paying into the entitlement systems. Use outside resources to determine what some European governments have done to combat this issue.

   ____
   ____
   ____

Exploration 8.3: EUROPE CULTURE

Turn on the *3-D Buildings* layer in the *Primary Database*, and then click the *La Defense* placemark in the CULTURE folder. Be patient when working with the *3-D Buildings* layer as it can take some time to download the data to your computer. Does this landscape match your ideas of the built environment of Paris? Now click the *Seine River* placemark, and study the landscape. Is this the Paris you had in mind?

Exploration 8.3: MULTIPLE CHOICE

1. What is the most distinct difference in the cultural landscape between the *La Defense* and *Seine River* views of Paris?

   A. the vertical development of the built environment
   B. traffic infrastructure
   C. the Seine River is a part of the landscape in one and not the other
   D. only one landscape has an iconic distinguishing architectural feature
   E. only one has a pedestrian oriented mall/boulevard

If you have visited Paris, you are familiar with the terms "Right Bank" and "Left Bank" in reference to the Seine River. If you are unfamiliar with this distinction, use outside resources to educate yourself about the meanings of these terms in Paris. Take note that the designations refer to more than just what side of the river you are on Paris. There are distinct cultural associations. For example, the Left Bank (la Rive Gauche) is thought of as the artistic side of the city, while the Right Bank (la Rive Droite) is home to the most affluent Parisians along with a high number of banks and businesses. From the initial view provided by the *Seine River* placemark, follow the river north and east until you reach le musée du Louvre.

2. Which of the following statements is correct in regard to the location of le musée du Louvre?

   A. It is located on the Ile de la Cité.
   B. It is located in la Defense.
   C. It is located on la Rive Gauche.
   D. It is located at Versailles.
   E. It is located on la Rive Droite.

Continue eastward on the stream until you reach another famous landmark, Notre Dame de Paris. This Roman Catholic cathedral in the Gothic style was one of the first buildings in the world to utilize a unique architectural feature. Be sure the *3D Buildings* layer is still turned on and study the building. The unique features are evident on the exterior of the building.

3. Notre Dame de Paris extensively utilizes

   A. a hypocaust.
   B. Doric columns.
   C. flying buttresses.
   D. Queen Anne-style spindles.
   E. Tidewater-style hipped roof.

Continue eastward on the stream for approximately six additional kilometers until you reach a fork in the stream. Approximately one kilometer east of the confluence point, there is a star-shaped structure. If you have some difficulty locating this structure it might be helpful to tilt view to vertical and zoom-out. This star-shaped feature is Fort de Charenton, built in 1842. Historic structures are plentiful and valued in Europe. They have often been preserved or restored. This historic structure has been repurposed many times since its construction. It is now utilized by the French National Police. Click the *Lyon* folder and you will be whisked to the south of France. The view of Lyon contains five labeled placemarks (1 to 5). One of these sites is the former location of a fort that was quite similar to the Fort de Charenton. Expand the *Lyon* folder and then click each of the numbered placemarks and attempt to locate the site that has vestiges of the old fort. A tool in the toolbar will provide you with an unquestionable answer if you want to be certain of your response.

4. The Lyon location of the former fort is site

A. 1.
B. 2.
C. 3.
D. 4.
E. 5.

The Europeans cherish tradition in their culture and one sporting tradition has a very special place in their cultural heritage. Le Tour de France is the most important cycling race in the world. There are actually two other Grand Tours, the Tour of Spain (Vuelta a España) and the Tour of Italy (Giro d'Italia), but le Tour is the biggest of them all. Globally, more people watch the Tour de France than any other annual sporting event. Le Tour lasts three weeks and covers approximately 3,500 kilometers in a combination of day-long segments called stages. The Mountain stages are often where Le Tour is won or lost. These grueling race days can last more than five hours with riders climbing one high mountain after another. Double-click the *Le Tour de France* folder and then expand it to see nine locations placemarked. Be sure the *Terrain* layer in the *Primary Database* is checked on. Beginning with *Pau* and progressing through the list to the *Col du Tourmalet*, follow the route of just one stage of the 2010 Tour de France.

5. Following the roads between the nine locations identified in the *Le Tour de France – Example Stage* folder, measuring and totaling the distance of this stage with the path option of the ruler tool. How long is this stage of the 2010 Tour de France?
A. 75 km
B. 100 km
C. 125 km
D. 175 km
E. 225 km

Exploration 8.3: SHORT ESSAY

1. Comment on the differences between the built environment in and around the La defense and Seine River placemarks. Beyond identifying the differences, detail what factors could explain the profound differences. Be sure to consider land value and public policy in your response.

____
____
____

2. Think about reasons why the Tour de France is such an enormously popular sporting event, particularly in Europe and France. Why is this sport more appealing to Europeans than Americans? Why would Le Tour have built up such a large following before the advent of television?

____
____
____

Exploration 8.4: EUROPE GEOPOLITICS

Verify that the *Terrain* layer is turned on, and the *3D Buildings* layer is turned on in the *Primary Database.* Europe is home to several microstates. A microstate is a sovereign state with very small populations and/or very small land area. Expand the GEOPOLITICS folder and then expand the *microstates* folder. You will see five European microstate placemarks. Click each placemark and examine the microstates.

Exploration 8.4: MULTIPLE CHOICE

1. Utilizing the information gleaned from your study of the microstates and any outside sources you find helpful, evaluate the following statements. Select the statement that is *not* accurate.

A. San Marino is an enclave.
B. Liechtenstein is sandwiched between Austria and Switzerland.
C. Malta is an island microstate.
D. Monaco is the least populous of the placemarked microstates.
E. Vatican City is the smallest microstate (in size).

Turn off the *3D Buildings* layer and turn on Historical Imagery (click the clock in the toolbar). Now double-click the *Berlin 1943* placemark. You will see an image from central Berlin in 1943. After you study the image, double-click *Berlin 1945* and note the changes to the city. Finally, double-click *Berlin 1953* to see how the city has changed over the next eight years.

2. In regard to the series of Berlin placemarks, which of the following statements is most accurate?

A. By 1953, Berlin had largely been rebuilt.
B. By 1943 there was widespread bombing damage in central Berlin.
C. All of the bridges in central Berlin had been destroyed in the 1945 image.
D. The Berlin Cathedral's dome had not been repaired by 1953.
E. There were significantly fewer structures in 1953 than in 1943.

Turn off the Historical Imagery and expand the *Berlin Wall* folder. Double-click the folder to zoom to the appropriate vantage point. The Berlin Wall was a prominent icon of the Cold War. This wall was a concrete barrier built by East Germany in 1961. This wall, which was often two walls with an open "zone of death" in between, completely surrounded the city of West Berlin until it was dismantled in 1989. Although the wall came down in 1989, one can still see numerous locations around the city that speak to the former presence of the barrier. A few sections of the wall remain for display, but most often the evidence is in the form of open space. Examine the five colored paths that are laid upon the imagery of central Berlin. One of these paths illustrates the location of the Wall through this part of the city. You will need to zoom in and study the paths to determine which one follows the old Berliner Mauer.

3. Which path follows the approximate route of the old Berlin Wall?

A. A (yellow)
B. B (orange)
C. C (green)
D. D (blue)
E. E (pink)

Conflict, and for that matter genocide, did not make their last appearance in the region with World War II. The disintegration of the political entity of Yugoslavia formally began with the independence of Croatia and Slovenia in 1991 and continues to this day. Many lives have been lost along the way as ethnic and national identities have wrestled for control of emerging states. One of many terrible episodes occurred with the Siege of Sarajevo. From April 5, 1992 until February 29, 1996 Serb forces encircled the city and assaulted the city with artillery, mortars, and sniper fire. Thousands were killed and wounded as hundreds of artillery shells fell on the city each day. This violence was a stark change from the theme of peace that filled the air in 1984 when Sarajevo hosted the Winter Olympics. Contained in the *Sarajevo* folder are three placemarks for Olympic sites. The *Igman ski jump* hill has a medal ceremony stand filled with bullet holes. The *Trebevic bobsled and luge track* was a popular site for Serbian artillery positions. If you turn-on the *Panoramio* images in the *Geographic Web* folder of the *Primary Database* you can view images of the war-torn track. When you click the *Zetra ice hall* placemark, you will see a legacy of the Siege on Sarajevo.

4. The landscape around the Zetra ice hall placemark shows a lasting impact of the Siege on Sarajevo. In this imagery one can see

A. the city is still in ruins.
B. that much of the open space around the Olympic facilities has been filled with huge cemeteries.
C. that the Olympic ice hall has not been repaired.
D. that there are still numerous tanks on the streets of Sarajevo.
E. there is no evidence of Olympic facilities left on the landscape.

The evolution of the political landscape of the former Yugoslavia continues to this day. The most recent issue involves Kosovo. Kosovo declared its independence in February of 2008. Be sure the *Borders and Labels* folder is checked on in the *Primary Database* and then double-click the *Kosovo* placemark. Do you see anything that suggests the border of Kosovo might be different in some way? Click the link associated with the *Kosovo* placemark and review the information on the international recognition of Kosovo.

5. With respect to Kosovo's declaration of independence, identify the place set that best explains why Spain, Russia, and China have not recognized Kosovo as independent from Serbia.

A. United States, France, Australia
B. Croatia, Slovenia, Bosnia and Herzegovina
C. Catalonia, Chechnya, Tibet
D. Iraq, Iran, North Korea
E. Albania, Montenegro, Macedonia

Exploration 8.4: SHORT ESSAY

1. Select one of the microstates highlighted in this exploration and describe the historical and geographic circumstances that led to the creation and ongoing viability of the state.

____
____
____

2. Do you think Kosovo should be recognized as an independent country? Explain your answer. Be sure to provide a comparative location/situation in your response.

___

___

___

Exploration 8.5: EUROPE ECONOMY AND DEVELOPMENT

Europe has been the source of many of the technological developments on which people from around the world depend. Two key threads of developments are transportation and astronomical knowledge. Open the ECONOMY AND DEVELOPMENT folder and click the *location A* placemark. You will see the labels for *location A* and *location B* on your screen. These places are significant because they are associated with one of the world's most important transportation innovations. Zoom in to the sites to help provide clues as to the significance of these locations.

Exploration 8.5: MULTIPLE CHOICE

1. Utilizing the location A and location B placemarks along with any outside resources, identify the innovator associated with these locations.

   A. Wehrner von Braun
   B. Louis Braille
   C. Rudolph Diesel
   D. Gerardus Mercator
   E. George Stephenson

Germany has had a prominent role in the initial and subsequent technology associated with a particular type of transportation technology. In the last century, this innovation had led to the development of an industry that represents a significant share of Germany's industrial output. There are six major producers of this technology in the country. One of the sites associated with this industry is associated with the *industry from innovation* placemark.

2. The site associated with the industry from innovation placemark produces

   A. locomotives.
   B. automobiles.
   C. aircraft.
   D. hovercraft.
   E. helicopters.

Within the European region the quest for new knowledge continues. One site of great interest is located along the French-Swiss border. This project is a remarkable example of international collaboration between thousands of scientists, universities and government entities from around the world. The project became operational in September of 2008. It is difficult to identify a spatial signature of the project via Google Earth™ as you can see by clicking the *science project* placemark. If, however, you click the *US example* placemark you will see a small-scale version of what is taking place in Europe.

3. Which of the following terms is most closely associated with the research taking place at the *science project* placemark?

A. pharmaceuticals
B. high-speed super train
C. extraterrestrial life
D. particle physics
E. plastics

Turn on the *3D Buildings* layer and click the *University of Pisa* placemark. The University of Pisa has numerous alumni of note. One of the most noteworthy in the realm of technological innovation is Galileo Galilei. Galileo's many contributions to science and technology include major improvements to the telescope. He is considered the father of modern observational astronomy. Without Galileo, the trajectory of the development of modern astronomical facilities could have been quite different. Now click the *Gran Telescopio Canarias* placemark. This is one of the largest telescopes in the world, located on Spain's Canary Islands. Let's take a look at what astronomers can see from this vantage point. Uncheck the *Europe* folder. In the toolbar, click the button with the planet icon and select "sky." Alternatively, you can click "View," then "Explore," then "sky." You will see that you now have a *Sky Database* folder in your list of layers. Expand the *Sky Database* folder and then expand the *Welcome to Sky* database. Turn on and then double-click the *Getting to Know Sky* placemark. Work through the brief tutorial so you will have a basic understanding of how Sky works.

4. When using Sky, which of the following options is not supported?

A. You can display a grid with red lines to orient yourself to latitude and longitude.
B. When you are viewing Earth in Google Earth™, you can switch to Sky at any time to view the sky from that point.
C. Historical sky maps are available.
D. Maps of extraterrestrial communications are available.
E. You can have Han Solo fly the Millennium Falcon across the sky.

Expand and turn on the Our Solar System layer. Examine the imagery provided for each planet.

5. Which planet displays visible ice at its south pole in the imagery associated with the Our Solar System folder?

A. Mercury
B. Venus
C. Mars
D. Jupiter
E. Saturn

Exploration 5: SHORT ANSWER

1. Research the activity that is taking place at the science project placemark. Briefly describe what the goals of the research at this site are. Also include a description of the concerns some people had with the instigation of research at this site.

____

____

____

2. Turn on *Constellations* layer in the *Backyard Astronomy* folder and the *Hevelius Overlay* in the *Hevelius Constellations* folder that is located in the *Historical Sky Maps* folder. Identify three constellations that have an artistic rendering associated with them. List the three constellations and describe the associated image.

   ____

   ____

   ____

YOU MAP IT! Tour de France Stage

Now that you have a feel for a Tour de France stage, let's see if you can design one yourself. Be sure the *Terrain* layer and the *Roads* layer is turned on. Go to the YOU MAP IT! folder and you will see *Start City* and *Finish City* placemarks. The Start City is Grenoble, France and the Finish City is Albertville, France. The highway distance between these two cities is approximately 80 kilometers. Your job is to identify a route that is between 125 kilometers and 175 kilometers long. You also want to incorporate as much climbing as possible in your route. You may find it useful to employ a web mapping service such as the Get Directions function of Google Maps. In this application you can drag the route to calculate different options.

Once you have identified the most suitable route, placemark five locations along the way. Create these placemarks in the YOU MAP IT! folder. Manipulate the view for each placemark so that it follows the path. Look back at the route in the Culture Exploration for an example. When you get your view oriented properly for each placemark so that you are pointed down the road, simply click "snapshot view" after right-clicking the placemark in the table of contents.

Create a document with the key information for your route. This will include the distance between each of the five placemarks along with an estimated elevation gain for each segment. Also provide the total distance and total estimated climbing for the entire route. The student who falls within the parameters of distance and incorporates the most climbing is the winner. You can ask your instructor if they will treat you to a trip to see Le Tour in person.

Turning it in:

Your instructor will provide you with an explanation of how to submit your results from this assignment. This may include e-mailing a .kmz file you create from your *YOU MAP IT!* folder and a document with the required information to your instructor.

To create a .kmz from your *YOU MAP IT!* folder, simply click once on the *YOU MAP IT!* folder to highlight it, then go to File, Save Place As…, and save it in an appropriate location on your computer.

Name: ______________________________

Date: ______________________________

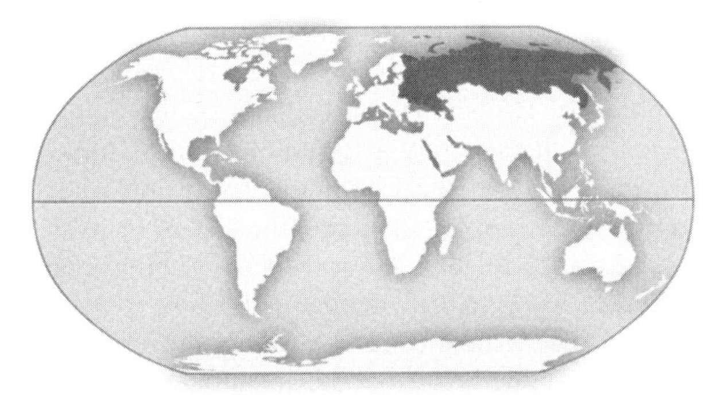

Chapter 9: The Russian Domain

The Russian Domain is a region that has experienced shifting boundaries in recent years. During the time of the Soviet Union, the Russian-centered state incorporated a number of Soviet Socialist Republics along its margins. Today, former Soviet Socialist Republics like Lithuania, Estonia, and Latvia are considered with the European region, while the "stan" states such as Kazakhstan and Uzbekistan are included with a Central Asia region. With the demise of the Soviet Union, the ethnic heart of the state constitutes the country of Russia. States that are under the strong influence of Russia and/or share cultural similarities are included in the Russian domain of today. The largest city is Moscow with over 11 million inhabitants.

The environment of the region is noteworthy for the toxic legacy of the Soviet Union. Two of the most acutely polluted areas are the industrial city of Norilsk and the zone surrounding the now-closed nuclear energy facility at Chernobyl, Ukraine. Today, new pollution is created by the millions of cars jamming Moscow streets, contributing to increasing levels of carbon dioxide in the atmosphere.

The Russian Domain is one of the few world regions where declining population presents problems. Birth rates remain low while higher death rates attributable to declining health combine to make this a problem. The region is predominantly urban. Very high population densities are found within cities at the mikrorayons, a quintessential Soviet settlement feature. Population in the central and eastern part of the region follows the transportation corridor of the Trans-Siberian Railroad.

The Russian Domain is a region that began a cultural transformation with the demise of the Soviet Union. In many ways the region looked westward as it increasingly incorporated cultural attributes of Europe and North America, like fast food. It also witnessed resurgence in long-suppressed cultural traditions such as the practice of Eastern Orthodox religious traditions.

The disintegration of the Soviet Union created some unique geopolitical challenges. For example, the Soviet Union had numerous nuclear weapons stationed in the now-independent states of Kazakhstan, Ukraine, and Belarus. While this issue was resolved, geopolitical tensions remain in the region as newer independent states and culturally distinct regions of Russia continue to define their roles and identities within the region.

The southwestern part of this region is not only a tense region due to ethnic and historical factors, but also for modern economic realities. This area is home to a number of energy corridors that move Russian oil and gas to European markets, and will remain relevant for the foreseeable future.

Download EncounterWRG_ch09_RussianDomain.kmz from **www.mygeoscienceplace.com** *and open in Google Earth™.*

Exploration 9.1: THE RUSSIAN DOMAIN ENVIRONMENT

When one considers the potential hazards of nuclear energy, one site stands above all others. The disaster that originated on April 26, 1986 at Chernobyl Nuclear Power Plant released enormous amounts of radioactive material in the surrounding environment. According to National Geographic, the radiation released was several hundred times more than that associated with the bombing of Hiroshima. Large parts of the Soviet Union and Europe were affected by the radioactive fallout. Most of the severely affected areas in Ukraine, Russia, and Belarus have now been deemed safe for occupation. However, the Chernobyl Exclusion Zone remains off-limits to settlement. Contained within this area are the former power plant and the city of Pripyat. Turn on the *3D Buildings* layer in the *Primary Database*, and then go to the *Chernobyl Power Plant* and *Pripyat* placemarks to survey the largely abandoned scene. Note: workers still maintain the power plant although it ceased power production in 2000. Read the hyperlinked BBC article about wildlife in the area and then turn on the *Panoramio Photos* in the *Primary Database* and sample the images of the abandoned city.

Exploration 9.1: MULTIPLE CHOICE

1. Which of the following statements is *not* accurate?

A. Fauna in the Chernobyl Exclusion Zone remains largely absent more than 25 years after the accident.
B. Flora has started to reclaim the urban environment of Pripyat.
C. The Chernobyl Power Plant has not been removed from the site.
D. Surface water is still present in the vicinity of Pripyat.
E. Birds are nesting inside the concrete sarcophagus erected over the exploded reactor.

Now click the *Zone of Exclusion* placemark and examine the wider landscape. Zoom in and out as necessary and attempt to determine the approximate radius of the Zone of Exclusion. You can look for areas that suggest they are currently inhabited and road blocks on routes leading to Pripyat.

2. The radius of the Zone of Exclusion around the Chernobyl Power Plant is approximately

A. 10 kilometers.
B. 20 kilometers.
C. 30 kilometers.
D. 40 kilometers.
E. 50 kilometers.

While the nuclear disaster at Chernobyl represents an acute case of pollution, there are locations in the Russian realm notorious for long-term pollution. One such locality is the Siberian city of Norilsk. Mining and excessive pollution has taken place in Norilsk for nearly 100 years. The area is known for its extensive deposits of nickel, palladium, and copper. The smelting of copper occurs on site and is responsible for up to one percent of the total global emissions of sulfur dioxide. Sulfur dioxide is a primary component of acid rain and subsequently the forests downwind (primarily south and east) of Norilsk are largely diseased or dead. Double-click the Norilsk folder and then open the link to NASA's Earth Observatory. Read the short interpretation of remotely-sensed imagery from Norilsk and then return to Google Earth™. Verify that the *Terrain* layer in the *Primary Database* is turned on. Expand the Norilsk folder and then tour the six Norilsk placemarks by clicking each one individually.

3. After viewing the Norilsk placemarks, which of the following statements is *not* accurate?

A. Surface water is visibly polluted.
B. All of the residents of Norilsk live in Soviet-era high-rise apartment buildings.
C. There are visible emissions from the numerous smoke stacks.
D. New landforms have been created from mine waste.
E. There is open-pit mining.

Pollution sources in the Russian Domain are not relegated to extreme point sources like Chernobyl and Norilsk. In fact, one area of rapidly growing carbon dioxide emissions is the city and suburbs of Moscow. The area of concern is the rapidly growing population of vehicles in Moscow. Many of the automobiles do not even have catalytic converters. The Russians have accommodated this boom in automobiles by a build-out of the boulevard and highway system in and around Moscow. Go to the Moscow placemark and turn on the *Roads* layer in the *Primary Database*.

4. Concerning the road network of Moscow, which of the following statements is *most* accurate?

A. They have many highways, but none with modern clover-leaf interchanges.
B. The outermost ring road has a circumference of 30 kilometers.
C. Moscow is served by one large north-south highway and one large east-west highway.
D. The main highways and boulevards of Moscow are organized in a right-angle gridiron pattern.
E. There are now no less than four ring roads around Moscow.

All the surplus carbon dioxide in our atmosphere contributes to the warming that we have seen in most global locations over the last century. One area in this region that is affected by warming temperatures is the Arctic Ocean. Turn on the Sea Ice Extent layer and then double-click the folder. You can follow the link to the National Snow and Ice Data Center to explore all kinds of interesting data in Google Earth™ related to the cryosphere. For now, just wait a few moments for the data to load then click the play button in the sea ice time-slider. The animation will need to repeat several times before the entire image set loads into the animation. Observe the patterns of sea ice extent from 1979 to 2008. When you complete this exploration, be sure to turn off the *Sea Ice Extent* layer.

5. Regarding sea ice extent from the period 1979 to 2008, which of the following statements is *least* accurate?

A. The minimum sea ice extent year of the dataset is 2007.
B. The maximum sea ice extent years occurred in the late 1970s and early 1980s.
C. The more dramatic declines in sea ice extent occurred following 2000.
D. The decline in sea ice extent has been greatest immediately north of Greenland.
E. More shipping can take place north of eastern Russia as a result of changing sea ice conditions.

Exploration 9.1: SHORT ESSAY

1. While there are many benefits from employing nuclear energy as part of the energy production matrix, there are also very small, yet unparalleled risks with using this technology. Research the Chernobyl nuclear disaster and nuclear power in general. To what extent do you believe nuclear power should be a part of the energy solution in your country?

___

___

___

2. The changing climate of our planet is creating vastly different conditions in some locations. While there are many losers with climate change scenarios, there can also be winners. Think about who would stand to benefit and why they would benefit if the sea ice extent in the Arctic Ocean continues its downward trend.

___

___

___

Exploration 9.2: THE RUSSIAN DOMAIN POPULATION

Declining populations in the Russian Domain represent the dominant demographic theme. Expand the POPULATION folder and turn on the *total fertility rate – 2008* folder. Study this layer and then turn off the *total fertility rate – 2008* folder and turn on the *RNI – 2008* folder. This layer illustrates the percentage increase or decrease for each country's population annually.

Exploration 9.2: MULTIPLE CHOICE

1. Based upon your analysis of total fertility rates and the rates of natural increase in 2008, which of the following statements about the region's countries is *most* accurate?

   A. Women in Ukraine are having the fewest children, but the population is declining most rapidly in Moldova.
   B. Women in Moldova are having the fewest children, and the population is declining most rapidly in Belarus.
   C. Women in Belarus are having the fewest children, but the population is declining most rapidly in Ukraine.
   D. Women in Armenia are having the fewest children, and the population is declining most rapidly in Belarus.
   E. Women in Russia are having the fewest children, and the population is declining most rapidly in Russia.

While the region's declining population is the most noteworthy demographic reality, the geographic spread of the population is an interesting facet as well. Turn off the *RNI – 2008* folder and turn-on the *population density* folder. You can see that the population of the region is distributed unequally across the region.

2. Which of the following statements is *least* accurate in reference to population distribution in the Russian Domain?

   A. The Russian Domain contains one of the largest expanses of sparsely populated land in the world.
   B. The higher density population corridor of the region is clustered around a major transportation route.
   C. The higher density population corridor is generally located in a band between 50° and 60° north latitude.
   D. Ukraine, Belarus, and Armenia generally have lower population densities than the country of Russia as a whole.
   E. The higher density population corridor of the region is clustered around a major transportation route.

Turn off the *population density* folder and click the *mikrorayons 1* placemark. Your view is a collection of mikrorayons. These are collections of high-rise apartment buildings that were primarily constructed during the Soviet-era. These centers are often clustered on the outskirts of major urban areas, like this one near Moscow. The mikrorayons are usually somewhat self-contained with services such as grocery stores and public transportation hubs. Zoom in and view this specialized urban landscape. Turn on the *Panoramio* images and sample some ground-level views.

3. Regarding the mikrorayons 1 urban landscape, which of the following statements is *not* accurate?

   A. There are recreational fields available.
   B. There are commuter rail lines nearby (less than 3 kilometers for most apartments).
   C. There is evidence of new construction amongst the mikrorayons.
   D. There is a major highway nearby (less than 5 kilometers for most apartments).
   E. These are large parking lots to accommodate multiple-vehicle families living in the apartments.

Now go to the *mikrorayons 2* placemark and complete a survey of that location. If you look closely around this site, you will see something in particular that potentially suggests a change in Russian values.

4. Which of the following features is visible at the mikrorayons 2 placemark that is more similar to what one might see in a North American urban landscape?

   A. The new construction of large single-family homes in a new housing development.
   B. A trend toward the abandonment of mass transit alternatives by the lack of a rail station.
   C. Decreased green space compared to mikrorayons 1.
   D. Significantly more parking available compared to mikrorayons 1.
   E. A large shopping mall in the center of the development.

The Trans-Siberian railroad is a network of railroads that connects Moscow to the Russian Far East. From its western terminal in Moscow to its eastern terminal in Vladivostok, the railroad stretches more than 9000 kilometers. The bulk of construction occurred in the late 19th and early 20th centuries. The railroad has played a huge role in the subsequent development of Russia's population and commerce. Open the link associated with the *eastern terminus* placemark. This will take you to a fascinating project by Google Russia that will enable you to watch videos, see pictures, and read descriptive text about the train along the route. Explore this website to get a feel for one of the world's great railroads. Eventually return to Google Earth™ and click the link associated with the *western terminus* of the Trans-Siberian. On the left-hand side of the screen, select Vladivostok, and then click the television icon on the map to watch a brief video about the city.

5. Based on information gleaned from the Google video on Vladivostok, which of the following statements is *not* accurate?

   A. You can purchase some nice right-hand drive vehicles in Vladivostok.
   B. The city's name is derived from a phrase that means "rule the east."
   C. Vladivostok is the site of Russia's defeat of Japan.
   D. There are massive artillery batteries in the hills above Vladivostok.
   E. Khrushchev promoted a rivalry with San Francisco's cable cars.

Exploration 9.2: SHORT ESSAY

1. Compare and contrast the *mikrorayons 1* and *mikrorayons 2* urban landscapes. Does one site appear to house more affluent residents? Detail at least three differences between the two locations in your discussion. In your discussion, keep in mind that these images were captured at different times of the year.

   ___
   ___
   ___

2. Select another one of the locations on Google Maps virtual journey from Moscow to Vladivostok aboard the Trans-Siberian Railroad. Identify the location and write a short paragraph that describes the site and highlights at least two cultural differences from what you would find in your community.

   ___
   ___
   ___

Exploration 9.3: THE RUSSIAN DOMAIN CULTURE

In 1990, the first McDonald's restaurant opened in the Soviet Union in Moscow's Gorky Park. You can see the location by going to the *Moscow's first McDonald's* placemark in the CULTURE folder. At the time, it was the world's largest and was a result of years of negotiations and tens of millions of dollars of investment by the western corporate behemoth. The restaurant still serves more than 30,000 persons a day. Today there are more than 125 McDonald's in Moscow and several dozen Russian cities with plans for many more. This is but one small example of the surge in globalization in the Russian Domain since the breakup of the Soviet Union. The menu of Russian McDonald's is quite similar to that of their counterparts in the United States with the addition of cabbage pies. McDonald's isn't the only US import in the fast-food realm. Search using the fly-to box to identify other American fast-food chains in the vicinity.

Exploration 9.3: MULTIPLE CHOICE

1. Which of the following fast-food chains does not currently have restaurants in the Moscow metropolitan area?

   A. Papa John's
   B. Burger King
   C. Arby's
   D. Pizza Hut
   E. Kentucky Fried Chicken

Russia is not the only location that the US-based McDonald's has expanded into in recent years. McDonald's now operates in more than 100 countries. In most of those countries, the menu reflects regional tastes. For example, go to the *McDonald's in Mumbai* placemark, and then click the link to the McDonald's menu. Think about an item that you would not expect to find at an Indian McDonald's and then check-out the menu available for McDelivery to verify your hypothesis.

2. Indian McDonald's do not serve

   A. hamburgers.
   B. French fries.
   C. chicken McNuggets.
   D. Filet-o-Fish.
   E. Coca-Cola.

The wide array of menu items that cater to regional tastes might surprise you. Open the *McDonald's around the world* folder. Zoom to each of the five placemarks and determine where you are located. Then utilize the link associated with *McD's A* to go to those various countries McDonald's websites and explore the menus.

3. Identify the McDonald's regional menu item that is not properly matched with its location.

   A. vegemite sandwich
   B. bacon potato pie
   C. McPollo
   D. McRib
   E. McArabia Kofta

Now that we have seen how cultural features have moved into the Russian Domain, let's think of a cultural characteristic that has diffused to other parts of the world. Let's begin by clicking the *Kremlin/Red Square* placemark. Verify that the *3D Buildings* layer is turned on and the *Panoramio* layer is turned off. From this viewpoint you can see several cathedrals, most notably Cathedral of the Archangel, St. Basil's Cathedral, Dormiton Cathedral, and Cathedral of the Annunciation. Perhaps you are surprised to find so many churches in the literal heart of Russia, knowing that the Soviet Union endorsed a policy of atheism. However, the Russian Orthodox Church is generally considered the largest of the Eastern Orthodox churches, with more than 135 million adherents worldwide. While the church and its clergy and followers were persecuted during the Soviet era, the institution has rebounded strongly over the last two decades. One unique aspect of many Russian Orthodox churches is the architecture of the church building. Examine the cathedrals and churches in the Kremlin/Red Square placemark to identify this distinctive architectural feature.

4. The distinctive architectural feature associated with Russian Orthodox churches is the

A. onion dome.
B. cantilevered steeple.
C. minaret.
D. flying buttress.
E. candy-striped prayer tower.

As the Russian Diaspora has spread around the world, a natural consequence is the concomitant spread of Russian Orthodoxy. Now that you have identified the distinguishing characteristic of Russian Orthodox churches by viewing the Kremlin/Red Square placemark, view the five placemarks contained in the *urban landscapes* folder and identify the location that has a Russian Orthodox place of worship. Because these locations have a high number of *3D Buildings*, it may take a minute or two for the scenes to load.

5. Studying the scenes contained in the urban landscapes folder reveals that there is a Russian Orthodox church in
A. Cape Town, South Africa.
B. New York City, USA.
C. Giza, Egypt.
D. Sydney, Australia.
E. Vienna, Austria.

Exploration 9.3: SHORT ESSAY

1. McDonald's is an archetypal American cultural export. What cultural exports do you associate with the Russian Domain? Complete some outside research and detail the diffusion of a cultural facet to the rest of the world.

____
____
____

2. Think of the religious landscapes in your community. Is there a dominant style or architectural feature(s) associated with the places of worship? Describe the similarities or common themes you have noticed in this regard.

____
____
____

Exploration 9.4: THE RUSSIAN DOMAIN GEOPOLITICS

When the Soviet Union dissolved in late 1991, a number of complex geopolitical questions had to be faced. One of the more serious was what to do with the network of Soviet weaponry that was spread across the newly independent states such as Ukraine. This country inherited more than 1000 nuclear warheads at the time of its independence. Over the period of several years these were transferred back to Russia and the old Soviet bases were largely closed or transferred to Ukrainian authorities. Ukraine signed the Nuclear Non-

Proliferation Treaty (NPT) and moved forward as a non-nuclear state. However, the legacy of the Soviet Union's nuclear past remains. Click on and then expand the GEOPOLITICS folder. Double-click the *Zoom to Soviet sub base* tour. There are two links associated with this tour. The first is the text of the NPT and the second is a link to a down-to-earth explanation of the treaty. Review these documents and then expand the *Balaklava* folder. Go to each of the five placemarks then do some internet searches to attempt to locate why this location is significant in Soviet military history.

Exploration 9.4: MULTIPLE CHOICE

1. Evaluate the placemarked locations around Balaklava and select the best response.

A. Site A was where the Soviet hovercraft fleet was based.
B. Site B was the port from which all of the Soviet military's balaclavas were shipped.
C. Site C was the home to the Soviet Union's Navy Special Forces unit.
D. Site D was the site of the Soviet Union's first nuclear test.
E. Site E was an entrance to a secret underground submarine base.

While Russia and the United States nuclear stockpiles are greatly diminished from the height of the Cold War, there are still thousands of warheads among these powers and the seven other countries that currently possess nuclear weapons. In fact, you can see plenty of evidence for Russia's continued nuclear virility in Google Earth™. Examine the nuclear sites beginning with the *type 1 site* placemark and working through to the *type 5 site* placemark.

2. After viewing the type 1 through type 5 sites, which of the following nuclear launch sites have you *not* seen?

A. Submarines
B. Rail-based launchers
C. Mobile launchers
D. Battlecruisers
E. Bomber aircraft

The distribution of nuclear weapons isn't the only geopolitical aspect that was shaken up by the dissolution of the Soviet Union. The transition from Soviet Union to Russia made for major shifts in borders, newly emerging countries, and some regions that sought independence but were not granted that status. The longest and bloodiest conflict associated with the lattermost situation occurred in Chechnya. The First Chechen War (1994-1996), Second Chechen War (1999-2000), and an ongoing low-level insurgency left this federal subject of Russia devastated. The capital city, Grozny, was destroyed in the first two wars. In 2003, the United Nations called Grozny the most destroyed city on Earth. What is the status of Grozny today? Go to the Grozny, Chechnya placemark and then navigate around the city to assess its condition.

3. Which of the following statements best describes the Chechen capital of Grozny?

A. The city is virtually destroyed, there are few intact structures.
B. Much of the city is heavily damaged, but signs of reconstruction are apparent.
C. New buildings are present, but no large-scale projects (e.g., stadiums or high-rise buildings) have been completed.
D. Much of the city has been rebuilt, ongoing construction is evident, but war damage is still visible.
E. The city appears to be completely rebuilt with no war damage visible.

Impacts of the Chechen conflicts have not been limited to the local environment. Two episodes tied to Chechen separatists standout. Events in Beslan and Moscow led to the deaths of many innocent people. You can click the *Moscow Theatre Siege* and *Beslan School Hostage Crisis* placemarks to see the locations of these terrorist acts. Turn-on the *Panoramio Photo* layer to view the images associated with these sites. Pay particular attention to the memorials associated with the locations.

4. The memorials of the Moscow Theatre Siege and the Beslan School Hostage Crisis portray a theme of

A. rising/flying.
B. revenge.
C. Russian political unity.
D. weeping/sorrow.
E. military victory.

Another conflict in the region flared recently when Georgia and Russia fought a brief war in August of 2008. The roots of this conflict reside in a debate about whether the parts of Georgia known as South Ossetia and Abkhazia should have been included with Georgia when it gained its independence in 1991 or if it should have remained a part of Russia. Turn on the *2008 South Ossetia war* image overlay. Study the map to gain an understanding of the geography of the conflict. Note the proximity of Grozny and Beslan. Also note the location of 2012 Winter Olympiad, Sochi Russia. When you have completed your analysis, be sure to turn off the *2008 South Ossetia war* image overlay.

5. With respect to the 2008 South Ossetia war, which of the following statements is least correct?

A. A naval engagement occurred in the conflict.
B. Russia attacked the Georgian capital city by air.
C. The Russian army occupied territory within 50 kilometers of the Georgian capital.
D. Georgia did not occupy any Russian territory.
E. No additional territory in Abkhazia and South Ossetia that had been loyal to the Georgian government was lost to the separatists.

Exploration 9.4: SHORT ESSAY

1. Research the state of nuclear armament in the world today. List the states that have nuclear weapons and what the general trends are in this arena. For example, are the numbers of states that possess weapons increasing or decreasing? Are the numbers of warheads in the world increasing or decreasing? What are the factors that steer the responses to these questions?

____
____
____

2. Research the 2008 South Ossetia war. In your opinion, was Georgia or Russia the aggressor? Why? Do you think the potential for more conflict remains? What would resolve the issue(s)?

____
____
____

Exploration 9.5: THE RUSSIAN DOMAIN ECONOMY AND DEVELOPMENT

The Russian Domain is a remarkably important region for the world's continued economic productivity. This lies in the fact that significant energy resources are located in the region. While energy is and will remain the region's economic focus, the Russian Domain contributes globally in other sectors of the economy. For example, wheat exports remain important. The history of wheat is a great story of globalization in and of itself as it was originally domesticated in the Fertile Crescent, but is now cultivated worldwide. The history of Russian/Ukrainian wheat, particularly a variety known as Turkey Red by North American farmers, is an essential part of the story of the rise of the US as a global economic power. This wheat, also known as winter wheat, is planted in the late fall and harvested in early summer across millions of acres in the US and Canada. Open the ECONOMY AND DEVELOPMENT folder and turn on the *percent arable land – 2008* folder. Examine the percentage of Russia that is arable in comparison to other countries around the world. You will find it useful to utilize your text and any outside resources that can provide you with information regarding climate regions of the world.

Exploration 9.5: MULTIPLE CHOICE

1. Select the statement that best describes Russia's percent of arable land status.

A. Russia has a higher percentage of arable land because it is the world's largest country.
B. In many parts of Russia, the climate is too cold and dry for many types of agriculture.
C. The Russian summers are too hot for agriculture to be successful.
D. Russia is a world leader in percentage of arable land because the climate is temperate throughout the country.
E. Russia's climate is generally too wet for many types of agriculture.

Russia's wheat exports may be the fourth largest in the world, but the Russian export that generates the most interest is energy in the form of oil and natural gas. Turn off the *percent arable land – 2008* folder and turn on the *oil exports- 2008* folder. Click the associated link to the US Energy Information Administration's (EIA) report on the status of oil development in Russia.

2. Select the statement that best summarizes Russian oil production and export.

A. Russian oil production has doubled in the last decade and today it is the world's leading exporter.
B. Russian production has climbed steadily from the early 1990s to the present and Russia is one of the top three oil exporters.
C. Russian production has been uneven over the last two decades but the country exported more oil than every country but Saudi Arabia in 2008.
D. Russian production has leveled off in recent years and it is no longer a top three oil exporter.
E. Russian production continues to climb steeply and they will continue to be a top exporter for the next 50 years.

Turn off the *oil exports- 2008* folder and turn on the *natural gas exports – 2008* folder. As you can clearly see, Russia was the world's leading natural gas exporter in 2008. Energy is *the* foundation for the contemporary Russian economy. Consumers will always need energy, but what they are willing to pay for it will vary depending on market conditions. Read the linked EIA report on Russian gas and then do an internet search for natural gas prices. Find a plot of natural gas and oil prices over the last five years.

3. Which of the following statements regarding the price considerations of oil and natural gas is most accurate?

A. Prices of these commodities are volatile and therefore revenues from exports will vary dramatically.
B. The price of natural gas continues to go up as supplies are exhausted.
C. As natural gas prices go up, oil prices go down. Therefore if you have both, you will always make money.
D. There is no relationship between oil and gas prices.
E. The price of natural gas continues to drop as renewable energy sources are more available.

Turn off the *natural gas exports – 2008* folder and turn on the *pipeline kilometers – 2008* folder. View the countries with the largest pipeline networks and then click the associated link to an EIA map of Russian pipelines. Assess the network of pipelines and the regions for future oil and gas development.

4. Select the statement regarding Russian oil and gas that is *not* supported by the map.

A. China is targeted as a major future consumer of Russian energy.
B. LNG export terminals have been proposed for northern, western, and eastern ports.
C. There are extensive regions that have been identified for future oil and gas development.
D. The Russian pipeline network is most developed along its eastern flank.
E. Russian oil and gas pipelines extend all the way to Western Europe.

Turn off the *pipeline kilometers – 2008* folder and go to the *Drohobych, Ukraine* placemark. Follow the associated link and read the EIA briefing.

5. Ukraine is important in the discussion of Russian oil and gas for all of the following reasons *except*

A. it does not have any oil and gas storage facilities.
B. it is one of Europe's largest energy consumers.
C. approximately half of its energy comes from natural gas sources.
D. it has access to warm water ports on the Black Sea.
E. its geographic position linking east and west.

Exploration 9.5: SHORT ANSWER

1. Review the oil and gas export folders and describe the geography of these energy export sectors. What regions are the exporting regions and what does this mean for future economic development for world regions?

____
____
____

2. Research contemporary events regarding the passage of Russian natural gas through the Ukraine. Provide a synopsis of thc situation. Be sure to address the overall implications for southern Europe and what has been suggested to mitigate future situations.

____

____

____

YOU MAP IT! Nuclear Armament

You have seen examples of Russia's nuclear arms site, but what about the rest of the world? Use outside resources to identify the states that are reported to have nuclear weapons. For each of these countries, search the web for nuclear locations like you viewed for Russia. When you locate a site, such as a launch pad or nuclear reactor used to weaponize uranium, placemark the location. Label the placemark with the country's name and in the description box of the placemark add information about the estimated number of nuclear warheads in the country's possession, the year of the first nuclear test for that country and the type of example site you have identified. Open the YOU MAP IT! folder to see an example for Russia.

Your instructor may want you to create a document with the information associated with your nuclear states.

Turning it in:

Your instructor will provide you with an explanation of how to submit your results from this assignment. This may include e-mailing a .kmz file you create from your *YOU MAP IT!* folder and a document with the required information to your instructor.

To create a .kmz from your *YOU MAP IT!* folder, simply click once on the *YOU MAP IT!* folder to highlight it, then go to File, Save Place As…, and save it in an appropriate location on your computer.

Name: ______________________________

Date: ______________________________

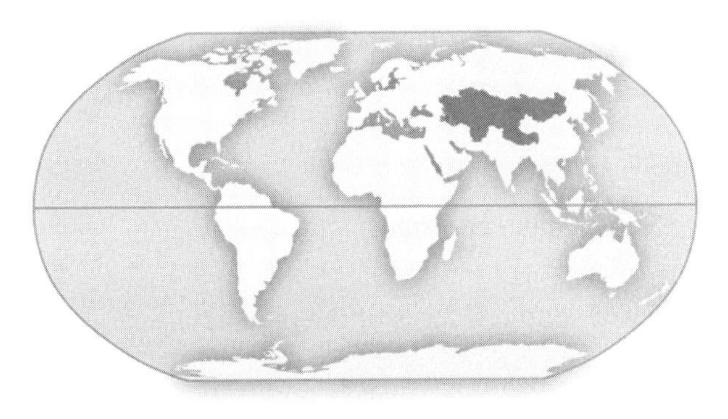

Chapter 10: Central Asia

The regional definition of central Asia is open to interpretation, and varied sources will define the region differently. For our sake, it includes five former Soviet republics: Kazakhstan, Kyrgyzstan, Uzbekistan, Tajikistan, and Turkmenistan, along with Azerbaijan, Mongolia, Afghanistan, and the Chinese provinces of Nei Mongolia and Qinghai, and finally the autonomous regions of western China known as Tibet and Xinjiang. It is generally a dry region that includes high mountains and a wealth of energy resources. Culturally, Islam and Turkic languages claim majorities in terms of religion and language. The region has historically been very rural, but many are now transitioning to urban lifestyles. Politically, the region has seen great unrest in recent years.

Central Asia is home to physical geography extremes. For example, the tallest mountain in the world is here. The region's arid environments have led to large-scale projects in human adaptation. These projects have not always led to the intended consequences. The Aral Sea is the most dramatic example of desiccation in the world, while the results of China's Great Green Wall are more ambiguous.

The people of Central Asia are distributed in a pattern that reflects the environmental realities of the region. This creates densely populated areas around hydrological resources. Complex geopolitical settings such as the Fergana Valley can result. Because of the region's pronounced aridity, people of Central Asia have developed an array of irrigation technologies to facilitate agricultural development.

Central Asia is a cultural crossroads in many ways. Altaic, Indo-European, and Tibeto-Burman language families come together here. This region also sees the eastern flank of Islam meet the eastern religious tradition of Tibetan Buddhism. Additionally, traditional Tibetan and Uygur cultures have waned in the face of ongoing Sinification as the Han Chinese populations increase in the region.

The region was reordered geopolitically when the Soviet Union dissolved in 1991. Since that time the newly independent states of Central Asia have often been viewed as lacking in the area of human rights. Human rights concerns can deal with inadequacies in social, political, or legal rights.

Not surprisingly, these challenges translate to a region that is also lacking in terms of social and economic development. Google Earth™ and related outside resources provide us with ample opportunities to explore the varied aspects of social and economic development. These explorations suggest that Central Asia and Afghanistan in particular, will need to make dramatic strides in development to approach the global median.

*Download EncounterWRG_ch10_Central Asia.kmz from **www.mygeoscienceplace.com** and open in Google Earth™.*

Exploration 10.1: CENTRAL ASIA ENVIRONMENT

The mountains of Central Asia are among the worlds highest. Mount Everest, also known as Chomolungma, is the highest mountain on Earth. Open the ENVIRONMENT FOLDER and go to the *Mt. Everest* placemark and explore the local environment.

Exploration 10.1: MULTIPLE CHOICE

1. Which of the following statements regarding Mount Everest is *not* accurate?

   A. Glaciers can be seen flowing from both the northern and southern faces of the mountain.
   B. It sits on the border between Nepal and China.
   C. The mountain is approximately 8848 meters high.
   D. The latitude of Everest is within a degree of that of Tampa, Florida.
   E. The closest large city (metropolitan population greater than one million) is Shigatse, Tibet.

Central Asia's mountains contribute to the region's marked aridity by blocking moist air from the interior parts of the region. The most extensive desert of the region is western China's Gobi Desert. Chinese citizens of the 20th century have witnessed the expansion of this dryland zone. To see firsthand, what encroaching dunes in the region look like; fly into the *dunes on the move* 360° panorama. The problem of encroaching dunes can present challenges for transportation infrastructure when roads and railroads are covered by the migrating dunes. Click the *dunes on the move 2* placemark and examine what else can be threatened by mobile dunes.

2. The dune field visible near the dunes on the move 2 placemark threatens

   A. a transportation corridor and agricultural fields.
   B. a transportation corridor and a surface water body.
   C. a surface water body and agricultural fields.
   D. an urban center and a surface water body.
   E. an urban center and agricultural fields.

All this sediment on the move can have impacts beyond the local advancing sand dune(s). Open the two links to NASA's Earth Observatory that are associated with the *dunes on the move 2* placemark. Review the images and the brief narratives provided by NASA.

3. Based on your assessment of NASA's Earth Observatory imagery and commentary on the Gobi Desert, select the statement that is *least* accurate.

   A. Dust storms from the Gobi can be hazardous to human health in terms of diminished air quality and reduced visibility.
   B. The largest Gobi Desert dust storms have a diameter of no more than 200 kilometers.
   C. Dust storms can result from factors such as deforestation, overgrazing, and drought.
   D. Dust from the Gobi is often described as yellow or yellow-tan in color.
   E. Dust storms from the Gobi can reach all the way to the United States.

The Chinese government has instituted a massive tree-planting initiative over the last several decades to slow erosion, protect water quality, improve air quality, and slow desertification. Open the *Great Green Wall* placemark. The Great Green Wall is an ongoing project that aims to plant more than 4,500 kilometers of forest strips. The linear tree rows have been planted in an attempt to block and/or slow the spread of dunes. This effort is one of the largest conservation efforts in world history. The results can be described as mixed, thus far. Other governments have attempted to deal with erosion problems with similar approaches. In the 1930s the United States planted a series of shelterbelts. Click the *shelterbelt* placemark to see a surviving remnant. You can click the link associated with the *shelterbelt* placemark to view an image of the types of events that precipitated the development of this adaptive strategy. Research the Great Plains shelterbelt system to learn more about this project.

4. Based on your outside research of shelterbelts and the shelterbelt placemark and link, identify the *least* correct statement.

A. Shelterbelts were promoted by US President Franklin Delano Roosevelt in the 1930s.
B. Shelterbelts seek to reduce wind velocity and decrease soil evaporation.
C. Shelterbelts were planted in a rectangular periphery around the Great Plains.
D. Government workers from the Works Progress Administration did much of the shelterbelt planting on the US Great Plains.
E. Millions of trees were planted on the US Great Plains during the Great Depression.

The Aral Sea is arguably the largest environmental crisis directly attributable to human actions that the world has ever seen. The Aral Sea was the world's fourth largest lake through the 1970's. However, Soviet irrigation projects diverted water from the Amu Darya and Syr Darya Rivers and the lake began to shrink. To witness the dramatic desiccation of the water body, open the Aral Sea folder. View the different image layers beginning with 1973 and working up to 2006 and then view the current Google Earth™ imagery by turning off the Aral Sea folder. Locate a recent image of the Aral Sea by searching the internet.

5. The Aral Sea is approximately

A. 1 percent of its 1973 level.
B. 10 percent of its 1973 level.
C. 25 percent of its 1973 level.
D. 50 percent of its 1973 level.
E. 75 percent of its 1973 level.

Exploration 10.1: SHORT ESSAY

1. Describe the economic impacts that are associated with the shrinking/disappearance of the Aral Sea.

___
___
___

2. Describe the health impacts that are associated with the shrinking/disappearance of the Aral Sea. Be sure to discuss the situation associated with the facility on Vozrozhdeniya Island. See the *Vozrozhdeniya facility* placemark.

___
___
___

Exploration 10.2: CENTRAL ASIA POPULATION

Central Asia is sparsely populated as a result of extreme environments. Many areas are too cold/high and/or too dry to support any significant populations. However, there are densely settled areas. Usually these locations are strongly tied to the availability of water. Expand the POPULATION folder. One of the most densely populated areas is known as the Fergana Valley. This fertile valley is watered by the Syr Darya River and its tributaries. The Fergana Valley is a complicated place because of ethnic diversity and issues of transnational water rights. Zoom to the *Fergana Valley* placemark. Be sure the *Terrain* layer in the *Primary Database* is turned on. Explore the Fergana Valley both upstream and downstream along the Syr Darya from the location of the *Fergana Valley* placemark.

Exploration 10.2: MULTIPLE CHOICE

1. From the point of the Fergana Valley placemark, what countries are encountered upstream and downstream along the Syr Darya?

A. Upstream: Tajikistan, Uzbekistan, Kazakhstan; Downstream: Tajikistan, Uzbekistan, Kyrgyzstan
B. Upstream: Tajikistan, Uzbekistan, Kazakhstan; Downstream: Tajikistan, Kyrgyzstan, Tajikistan
C. Upstream: Tajikistan, Uzbekistan, Kazakhstan; Downstream: Tajikistan, Uzbekistan, Kyrgyzstan
D. Upstream: Uzbekistan, Tajikistan, Turkmenistan; Downstream: Tajikistan, Kyrgyzstan, China
E. Upstream: Tajikistan, Uzbekistan, Kyrgyzstan; Downstream: Tajikistan, Uzbekistan, Kazakhstan

The next five placemarks are located along the periphery of the Tarim Basin. The Tarim Basin is a closed drainage basin. In other words, water drains into this basin and evaporates instead of making its way to the sea. The Tarim Basin is located in the western Chinese autonomous region of Xinjiang. The largest group that inhabits this area is known as the Uygur. Settlements such as those placemarked in the Tarim Basin have developed unique systems of livelihood. Explore these settlements now.

2. It appears that the five Tarim Basin communities emphasize an economy based on

A. petroleum and natural gas.
B. transportation.
C. manufacturing.
D. dryland (rain-fed) farming.
E. irrigated agriculture.

Expand your exploration to the wider Tarim Basin and Taklimakan Desert, paying attention to the patterns of settlement and associated infrastructure (highway networks, railroads, airports).

3. Based on your assessment of the five Tarim communities and the surrounding Tarim Basin and Taklimakan Desert, what statement best characterizes the population distribution of this part of Central Asia?

A. Evenly distributed across the Tarim Basin.
B. Disproportionately clustered into one major city.
C. Dispersed regularly along the super-highway that runs east-west through the Tarim Basin.
D. Clustered in small communities around the periphery of the Tarim Basin.
E. Disproportionately clustered along Trans-Tarim Pendi Railroad.

Now go to the *karez* placemark. What are these strange lines? Manipulate your view and utilize the clue that you have been given in this placemark's title to develop an understanding of what you are seeing.

4. Which of the following statements most accurately describes the lines in the karez placemark?

A. This is one result of China's Cultural Revolution.
B. The purpose of these lines has not been positively ascertained.
C. This is part of a new telecommunications network.
D. This is a tunnel system that is part of China's defense program.
E. This is an ancient stopover along the Silk Road because of the resource provided by these lines.

Moving forward to the *connect the dots 1* and the *connect the dots 2* placemarks; interpret these unique features of the settlement landscape.

5. Based on your study of the cultural landscape and any outside information, identify the term that best describes the features illustrated in the connect the dots 1 and connect the dots 2 placemarks.

A. qanat
B. tazza
C. adobe
D. azimuth
E. nadir

Exploration 10.2: SHORT ESSAY

1. Review the five Tarim Basin placemarks again. What shape does the area developed for agriculture most resemble? Describe what causes this characteristic spatial signature.

____
____
____

2. In this exploration, you have seen at least two different ways that communities have maximized local resources through the development of irrigation technology. Describe an example of water management in your community. What kinds of technology are required to make the system work? What benefits does it provide?

____
____
____

Exploration 10.3: CENTRAL ASIA CULTURE

While Central Asia has a high degree of physical homogeneity, the same cannot be said for the cultural realm. Clashes of culture have occurred throughout recorded history as the region has been a trade thoroughfare for thousands of years. Today, there continue to be cultural strains; some are acute and some are more generalized. A cultural clash occurred in the Bamiyan Valley of Afghanistan in 2001. At the time, Afghanistan was controlled by the Taliban, an Islamist organization with little tolerance for outside influence.

Go to the *Bamiyan Valley* placemark. Research the site and view the *Panoramio* images available in the *Geographic Web* folder of the *Primary Database*.

Exploration 10.3: MULTIPLE CHOICE

1. What occurred at Bamiyan in 2001?

 A. The Taliban expelled western aid workers.
 B. The Taliban constructed Islamic monuments.
 C. The Taliban attacked Christian missionary groups.
 D. The Taliban destroyed ancient Buddhist sculptures.
 E. The Taliban provided an Islamic sanction for Muslim poppy farmers.

Some people of the region have been strongly opposed to western influences on the cultural traditions of the region. Some of these fundamentalist groups believe it is an abomination to their cultural and religious traditions for outsiders to dilute their culture with divergent foods, music, language, and attitudes toward women for example. While some have argued against western influence with constructive dialogues, other have resorted to violence. Examine the *Baku* and *Kandahar* placemarks. Study the activities that are taking place in and around these two locations. If necessary, utilize outside resources to determine what external influences are dominant in Baku, Azerbaijan and Kandahar, Afghanistan.

2. Baku and Kandahar have a surplus of western cultural influence because they are respectively hubs for

 A. international oil and gas operations and the foreign military.
 B. whaling and soybean farming.
 C. commercial aircraft and the computer chip industry.
 D. higher education and automobile manufacturing.
 E. space technology and medicine.

Tibet is a fascinating cultural region with a very complicated geopolitical history. Today it is an autonomous region of China. One reason Tibet has remained a high level of cultural distinction throughout history is its rugged and isolated location. Tibet is the highest region in the world, with an average elevation over 4500 meters. The numerous mountains have contributed to the difficulty in accessing the region. As transportation technology has improved and the Chinese authorities have encouraged migration to the region, Tibet has become less isolated. In the last decade, a dramatic leap forward was made in terms of access to Tibet. Open the *Qinzang* placemark and survey the landscape. You can see the new addition.

3. Tibet recently became more connected to the rest of China with the opening of

 A. a six-lane super highway.
 B. a large inland port to facilitate barge traffic.
 C. a railway.
 D. an international airport.
 E. a multi-dish satellite communications center.

Western China is a region undergoing cultural transformation. One can see a spectrum of cultural landscapes from those that represent the older traditional ways to the new Han Chinese urban landscapes. Fly into the *Shannan A* and *Shannan B* panoramic images for comparative scenes in Tibet.

4. Based on your assessment of the Shannan A and Shannan B images, which of the following statements is *least* accurate.

   A. Within Shannan, there are still traditional Tibetan cultural landscapes.
   B. Modern conveniences and improvements are more evident in the Shannan A cultural landscape.
   C. Shannan A is a more distinctive/unique cultural landscape than Shannan B.
   D. Within Shannan, there are cultural landscapes that incorporate classical western architecture.
   E. Shannan A and Shannan B can both be considered symbolic cultural landscapes.

Nearby is the city of Lhasa, the capital of the Tibetan Autonomous Region in China. Lhasa is considered the birthplace of Tibetan Buddhism and was the seat of the Dalai Lama before he fled to India in 1959. The Potala Palace was the residence of the Dalai Lama for several hundred years and is thus an important cultural symbol. The Chinese government has turned the building into a museum and the location is a UNSECO World Heritage site. Let's examine two different perspectives of the grounds surrounding the Potala Palace. Study the differences between the *Potala A* and *Potala B* 360° panoramic views.

5. Identify the *incorrect* statement based on your assessment of the Potala A and Potala B panoramic views.

   A. Potala A reflects a more traditional Tibetan cultural landscape.
   B. Potala B reflects a more modern Chinese cultural landscape.
   C. Potala A emphasizes religious elements as evidenced by the misbaha wheels and sikharas.
   D. Potala B emphasizes Chinese nationalism as emphasized by flags and state monuments.
   E. Potala A and B present very different perspectives on cultural identity for Tibet.

Exploration 10.3: SHORT ESSAY

1. What do you think should be done with the Bamiyan Valley? Can and should the site be restored? What would be the benefits of restoration? What arguments are there for leaving the site as it is?

   ____
   ____
   ____

2. When you view the Potala B panoramic photo, you see some large Chinese characters. A loose translation of these characters is "Enthusiastically celebrating 70 years of the Chinese Communist Party." How do you think this message is received? What about its placement in front of an icon of Tibetan nationalism?

   ____
   ____
   ____

Exploration 10.4: CENTRAL ASIA GEOPOLITICS

Freedom House is a global leader in the study of human rights and political freedoms. Its annual reports evaluate the level of democratic freedom around the world. Let's evaluate the state of freedom in Central Asia in 2009. Open the GEOPOLITICS folder and then open the associated link to reach the Freedom House 2009 report. Use the map and/or drop-down box to assess the Political Rights and Civil Liberties scores for the countries of the region. The Freedom House rankings system goes from one (most free) to seven (least free).

Exploration 10.4: MULTIPLE CHOICE

1. In 2008, only eight countries in the world had average scores (political rights and civil liberties) of 7.0. How many of these worst of the worst countries were located in Central Asia?

A. 0
B. 1
C. 2
D. 3
E. 4

2. Freedom House defines countries with an average freedom score of 5.5 or higher as "not free." How many Central Asian countries (Afghanistan, Azerbaijan, China, Kazakhstan, Kyrgyzstan, Mongolia, Uzbekistan, Tajikistan, and Turkmenistan) fall into this category?

A. 1
B. 3
C. 5
D. 7
E. 9

Central Asia also fares poorly in a global comparison of another measure of freedom, that of the press. Freedom of the press refers to the ability to communicate through various media without state interference. In Central Asia there is a long history of state governments influencing the press through strict control of information and/or intimidation and imprisonment of journalists. View the selected Central Asian capital cities (the seats of government) and click on the associated links for each location. These links will take you to Reporters Without Borders briefings on the state of the free press in Central Asia.

3. Which of the following statements is not supported in your review of freedom of the press in the Central Asia region?

A. In Afghanistan, the Taliban have attacked the press and abducted journalists.
B. Uzbekistan's Islam Karimov has worked diligently to eliminate the independent press.
C. Azerbaijan represents a rare exception in Central Asia where freedom of the press has sharply increased in recent years.
D. Journalists and opposition figures have been assaulted and/or killed in Kyrgyzstan as recently as 2009.
E. Citizens of Turkmenistan have only been allowed to have individual connections to the internet since 2008.

One Central Asian country has the most notorious record when it comes to a free press. Complete control of information is a goal of this government and subsequently it consistently ranks as one of the worst countries for a free press in Reporters Without Borders annual rankings. You can see those rankings by following the link associated with the *TV broadcast stations – 2008* folder. Not surprisingly, this country only has a handful of broadcast stations. This enables the government to better control the information that is distributed to its citizens. View the *TV broadcast stations – 2008* data by turning on the folder.

4. What central Asian country was included in Reporters Without Borders bottom three countries in 2007, 2008, and 2009?

A. Afghanistan
B. Azerbaijan
C. Tajikistan
D. Turkmenistan
E. Uzbekistan

Turn off the *TV broadcast stations – 2008* folder. If Central Asia is the world region with the least press freedom, what are places with the most? You might be surprised to find affluent western democracies like France and Italy slipping down the rankings in recent years. In 2009, France had fallen to 43rd and Italy 49th. The United States just returned to the top 20 in 2009. It had dropped as low as 53rd in 2006.

5. What region of the world demonstrated the highest levels of press freedom in 2009?

A. Africa
B. Asia
C. Europe
D. Latin America
E. North America

Exploration 10.4: SHORT ESSAY

1. What could explain Central Asia's poor performance in terms of freedom and human rights?

____
____
____

2. Evaluate and describe the trend of press freedom in your country. What are the kinds of events and policies that affect the rankings in stable, affluent democracies where journalists are unlikely to be kidnapped?

____
____
____

Exploration 10.5: CENTRAL ASIA ECONOMY AND DEVELOPMENT

Central Asia is a laggard region in terms of economic development. Social conditions in terms of education and health in the former Soviet Socialist Republics have remained relatively better, but have begun to deteriorate in many cases. These characteristics can be challenging to study from the imagery of Google Earth™ alone. Therefore, it's a good idea to be aware of and make use of the multitude of resources provided by Google Earth™ users, news organizations, governments, non-governmental organizations, and corporations. There are a number of sites that aggregate this content. These include Google Earth™ Blog,

Google Earth™ Community, and Google Earth™ Hacks. For example, the Canadian Broadcasting Corporation (CBC) produced an outstanding six-minute Google Earth™ tour that addresses social and economic issues in Afghanistan. Open the ECONOMY AND DEVELOPMENT folder and then open the link associated with the *Afghanistan* placemark. This will open a blog entry from Google Earth™ blog. Click the "watch it here" link to open the CBC webpage. It's possible you will need to install the Google Earth™ plug-in to view the tour. A link is provided on the page. Watch the tour and answer the subsequent questions.

Exploration 10.5: MULTIPLE CHOICE

1. Which of the following statements regarding the economic and social development of Afghanistan is *not* supported by the Google Earth™ tour?

A. Afghanistan's arable land is less than 15% of the country.
B. As a result of increased trade with Iran, Herat is a relatively more prosperous and secure Afghan city.
C. More than seven million people in Afghanistan do not have enough food to meet their daily needs.
D. Clean water and electricity has been made available for nearly the entire Afghan population.
E. Afghanistan has an average per capita income of less than $1000 a year.

2. Afghan farmers and the Afghan economy disproportionately rely on the cultivation of

A. millet.
B. oats.
C. poppies.
D. rice.
E. wheat.

Perhaps you are interested in learning more about the cultivation of Afghanistan's chief crop. The various Google Earth™ blogs would be a great place to start. Let's try the Google Earth™ Community. In your web browser go to the homepage for the Google Earth™ Community (bbs.keyhole.com). Type "Afghanistan agriculture" into the search box and select the result that addresses the commodity you identified from the previous question (it will likely be the first result). Open the link and read the discussion and view the images regarding this crop. Keep in mind this is not an academic site, it is content provided by everyday people. When you reach the bottom of the blog entry, you will see a kmz file that begins with the numbers 1003165. Click the link and then select "open" in the file download box. Google Earth™ will fly you to an agricultural landscape where the commodity in question is grown extensively. The creator of this kmz has added some interesting placemarks for you to gain more insight into the cultural landscape. You can expand the folder that will have been loaded into your *Temporary Places* folder to read descriptions of each placemark.

3. Which of the following statements cannot be verified by the content obtained from Google Earth™ Community?

A. The Helmand River valley in Afghanistan is intensively cultivated.
B. Canals and qanats are utilized extensively for cultivation in Helmand province.
C. The Bost Fortress is being utilized as a storehouse for the majority of agricultural production in the region.
D. The Boghra irrigation canal marks the western extent of cultivation in northern Helmand province.
E. There is evidence of local trade provided by one of the placemarks.

Open the link associated with the *Central Asia economic freedom* placemark. This will open a webpage that provides information regarding measures of economic freedom. It should be noted that this website is a creation of the Wall Street Journal and the Heritage Foundation. As with any source of data and/or information, you are strongly encouraged to familiarize yourself with the source and evaluate its objectivity. Regardless of your opinions of the Heritage Foundation, this site can be a useful tool to make regional and country comparisons of assorted economic criteria.

4. In terms of the index of economic freedom, what statement does *not* describe the conditions in Central Asia?

A. While lower on the economic freedom index, all of the countries in Central Asia are showing improvement.
B. The majority of countries in the region have levels of economic freedom that are described as repressed or mostly unfree.
C. There are extensive regions that have been identified for future oil and gas development.
D. Not a single Central Asian country registers as free or mostly free on the index of economic freedom.
E. The highest ranked Central Asian countries are only moderately free.

5. Identify the Central Asian country with the lowest economic freedom score, and then click the country's name to link to a more detailed report of economic freedom conditions within that country. Identify the statement that is not supported by the data. Be sure to really explore the data by reading exactly what each economic freedom measure means according to the Heritage Foundation.

A. Government spending is much lower than the global average.
B. Business freedom is much lower than the global average.
C. Property rights are implemented arbitrarily or not at all.
D. Foreign participation in the economy is very restricted.
E. Corruption is pervasive.

Exploration 10.5: SHORT ANSWER

1. Discuss the dilemma associated with agriculture in Afghanistan. Should farmers be able to cultivate the crop that provides the most income for their families?

____

____

____

2. Research the Heritage Foundation online to determine their ideological perspective. How might this perspective influence their representation of global economic issues?

____

____

____

YOU MAP IT! Seven Summits

We visited the world's highest mountain, Mount Everest, also known as Chomolungma. Open and turn on the YOU MAP IT! folder to view the placemark associated with this peak. You will see information has been entered in the placemark's Properties box that documents the local name of the peak, the country in which it is located, the elevation of the peak, and the year and persons involved in the first ascent.

Create similar placemarks for the high points on the other six continents. Be sure the *Terrain* layer is on, and create snapshot views of the peaks that provide perspective, like the Mount Everest example.

Your instructor may want you to create a document with the information associated with your seven summits.

Turning it in:

Your instructor will provide you with an explanation of how to submit your results from this assignment. This may include e-mailing a .kmz file you create from your *YOU MAP IT!* folder and a document with the required information to your instructor.

To create a .kmz from your *YOU MAP IT!* folder, simply click once on the *YOU MAP IT!* folder to highlight it, then go to File, Save Place As…, and save it in an appropriate location on your computer.

Name: ______________________________

Date: ______________________________

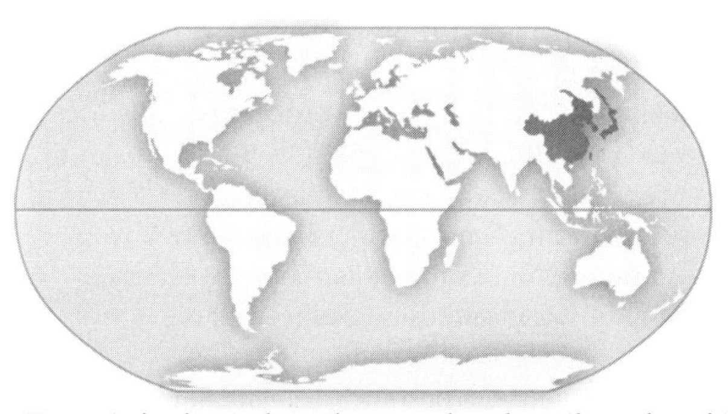

Chapter 11: East Asia

East Asia is a densely populated region that is dominated by the cultural, economic, and political influence of China. Japan, North Korea, South Korea, and Taiwan are also included in the region. China has a population of more than one billion, but the largest urban concentration is the greater Tokyo metropolitan area of Japan, where more than 30 million people live.

The physical environment of the region is diverse with high mountains, semi-arid grasslands, and extensive forests. Climatologically, comparisons can be drawn to North America. The region faces a host of environmental issues that are largely tied to China's explosive growth. Severe air pollution tied to China's reliance on coal for energy production and lax environmental regulations in industry combined with ongoing soil erosion along its major watersheds have forced China to strengthen environmental policies. Building the Three Gorges Dam attempts to mitigate some of these problems, but also brings new concerns.

The population of the region exhibits interesting patterns across space, time, and demographic groups. China has both significant rural populations and rapidly growing megacities.

The imprint of Chinese civilization on the region cannot be overstated. Distinctive Chinese culture emerged more than 4,000 years ago and is particularly distinctive because it evolved relatively independently from the world's other major culture hearths for some time. The most significant religious traditions of the region are Confucianism and Mahayana Buddhism while many of the languages have been derived from or strongly influenced by Classical Chinese.

The Chinese influence in the region is not limited to cultural traits. The geopolitical imprint of the country is large but its territory and boundaries are not universally accepted. On the other hand, a very clearly defined boundary is associated with the division of North and South Korea. The world's most prominent relic boundary exists in the region with the Great Wall of China.

The Great Wall is a fascinating component of the region's geopolitical history, but it continues to impact the region as a tourist destination. Preservation and development of this resource has been uneven along its path. While tourism is a key economic driver in the region, East Asia's export business has been the focal point of economic development for the last half century. Whether it's the array of products from China and Taiwan, automobiles from Japan, or electronics from South Korea, the region's exports are part of everyday life for many people around the world. In fact, the flood of exports has created significant trade deficits for many countries outside the region. This phenomenon has been particularly evident with the United States in recent decades.

Download EncounterWRG_ch11_EastAsia.kmz from ***www.mygeoscienceplace.com*** *and open in Google Earth™.*

Exploration 11.1: EAST ASIA ENVIRONMENT

China has experienced rapid economic growth in recent decades. The push for increased economic output has left its imprint on the Chinese landscape. A conspicuous example is the extensive erosion that has taken place on the Loess Plateau of China. The Loess Plateau is centered on the middle and upper reaches of the Huang He River. It gets its name from the fine silt that has been deposited by aeolian (wind) processes. This loess-based soil is highly prone to erosion when it lacks cover due to agricultural overexploitation, deforestation, and overgrazing. Open the ENVIRONMENT folder and explore the three soil erosion placemarks. You will see the characteristic yellow soils of the Loess Plateau. With every passing year, farmers lose more and more of their farmland to erosion. One strategy farmers can use to battle erosion is the construction of terraces. A terrace is a leveled area of a hillside that is designed to slow runoff of water and thus reduce erosion.

Exploration 11.1: MULTIPLE CHOICE

1. Is there evidence of terracing in the vicinity of the soil erosion placemarks?

A. No locations show evidence of terraced fields.
B. Soil erosion 1 shows evidence of terraced fields.
C. Soil erosion 2 shows evidence of terraced fields.
D. Soil erosion 3 shows evidence of terraced fields.
E. All locations show evidence of terraced fields.

2. Select the statement that is not verified by the area around the soil erosion 3 placemark.

A. This area is part of the Huang He watershed.
B. Erosion is limited to the east side of the floodplain in this vicinity.
C. Human settlements appear to be in the direct path of some erosional gullies.
D. Transportation networks have been affected by erosion.
E. Erosion is readily evident upstream and downstream of this location for more than 30 kilometers.

Now fly into the *Yungang Grottoes* and the *Linfen Yu King City Ruins* panoramic images. When you enter the Yungang Grottoes image and zoom forward into the wall, you can getter a better feel for the loess as you can see the individual strata or layers that were deposited by the wind over the millennia. Both of these sites illustrate how humans can adapt to their natural soundings and make use of the resources present.

3. The Yungang Grottoes and the Linfen Yu King City Ruins panoramic images illustrate that

A. people of the Loess Plateau have utilized rock-cut architecture.
B. everyone in the Loess Plateau is religious.
C. no vegetation grows on the Loess Plateau due to soil erosion.
D. there is not an opportunity for tourism in a land with severe erosion.
E. the Loess Plateau is uninhabitable.

With a large region of China identified as the Loess Plateau, and remarkable erosion occurring there, the sediment has to go somewhere. That somewhere for much of the Loess Plateau is the Huang He River, also known as the Yellow River for its sediment-laden waters. Go to *Huang He Delta, China* folder and view the *1979* layer. Then view the *2000* layer, and finally turn off the overlays to examine the most recent imagery from Google Earth™. You are witnessing the expansion of a particular physical geography feature as a result of the high rates of erosion along the Huang He and the subsequent deposition of material at the end of the stream. What's most remarkable is the time span at which this is happening. Anthropogenic erosion has greatly accelerated the process.

4. The feature being formed at the end of the Huang He is known as a(n)

   A. abyssal plain.
   B. archipelago.
   C. canyon.
   D. delta.
   E. esker.

While erosion has been an ongoing problem in China that is now being addressed through modified land use and reforestation, the problems associated with an exploding thirst for energy continue to grow rapidly. This is primarily reflected in China's reliance on coal for energy production, with more than two-thirds of its energy coming from this source. Therefore, China's coal-fired electrical generation plants represent a significant percentage of the world's CO_2 emissions (at least 15%). Zoom to the *coal-fired plants* placemark to view the largest of China's CO_2-producing plants. The Zouxian power facility produces more than 33 million tons of carbon dioxide annually. Beyond carbon dioxide, coal-fired plants also particulate matter, nitrous oxides, sulfur dioxide, various acidic compounds, and radioactive trace elements. In other words, they have the potential to seriously degrade air quality. This became a major issue in 2008 when a major event took place in Beijing. Go to the *clean air's important* placemark, turn on the *3-D Buildings* layer in the *Primary Database* and examine the scene.

5. The 2008 event associated with the clean air's important placemark was the

   A. summer Olympics.
   B. World Trade Organization meeting.
   C. G-20 meeting.
   D. Tiananmen Square anniversary.
   E. World's Fair.

Exploration 11.1: SHORT ESSAY

1. China has abundant coal resources that are being utilized to propel their economy forward. However, the Chinese are contributing enormous amounts of carbon dioxide to the atmosphere in the process. Do you think that the Chinese should be held to the same environmental standards as developed regions like North America and Europe or that they should be allowed the flexibility to pursue development with fossil fuels, just as the West did?

   ____
   ____
   ____

2. Explore the Three Gorges Dam folder and study the 1987, 2004, and contemporary images of the region. Utilize outside resources to research the controversies associated with the construction of this project and write a summary paragraph.

   ____
   ____
   ____

Exploration 11.2: EAST ASIA POPULATION

East Asia is home to one of the world's three great population clusters, but this population is not evenly distributed across the region or within countries. This is most striking when examining the range of population densities found between the rural regions of western China and the largest cities of the east. Open the POPULATION folder and turn on the *total population – 2008* folder. It's easy to see that the population of China is the largest in the world. This is good for comparisons between countries, but it doesn't shed light on where the people are located within the country. Turn off the *total population – 2008* folder and turn on the *population density* layer. This layer gives us a much better idea of the distribution but it is difficult to gauge the overall totals. This is just one very brief example of the value in portraying your data in different ways.

Exploration 11.2: MULTIPLE CHOICE

1. Based on the total population – 2008 folder and the population density layer, identify the statement that is *not* supported by the data.

A. Shandong and Jiangsu provinces are two of China's more densely populated provinces.
B. Eastern provinces of Taiwan are more densely populated than western provinces.
C. South Korea's population is nearly twice that of North Korea.
D. North Korea and South Korea have relatively similar population distributions.
E. Eastern China is much more densely populated than western China.

Expand the *Gallery* folder in the *Primary Database*. Locate and expand the *NASA* folder, and then check the box next to *Earth City Lights*. This is a large data set, so it takes a little while to load. Study the East Asia region and identify patterns in the lights. Where are the greatest clusters? Where is it darkest? Is this purely a reflection of population, or do levels of development help explain the patterns, as well? Think about the patterns of the *population density* layer. Compare the two layers and identify the most notable discrepancy between population density and light density/intensity.

2. Using the Earth City Lights and population density layers as your guide, identify the area of East Asia that has lower light output than expected.

A. North Korea
B. the Shanghai metropolitan area
C. the Tokyo-Yokohama metropolitan area
D. Japan's Kyushi Island
E. South Korea

Turn off the *Earth City Lights* and *population density* layers. Fly into the *Hong Kong* image to view one of the more densely populated landscapes of the world. Hong Kong's population density exceeds 6,000 persons per square kilometer. As you can see many of the buildings in Hong Kong have a vertical component to them. This is a reflection of high property values. As one moves away from vertical Hong Kong into adjacent areas, the urban form transitions to more horizontal buildings and fewer skyscrapers. Of course this leads to more space being required to house the same number of persons or businesses. Shenzhen is located just across from Hong Kong and has witnessed rapid urbanization in recent decades. Expand the *Shenzhen, China* folder and then turn on and zoom to the *1979* layer. Be sure the view is not obscured by folders/layers in the *Primary Database*, such as the Panoramio images. If so, turn those off. Proceed with viewing the subsequent images from *1990*, *2000*, *2004*, and the contemporary imagery from Google Earth™. The green areas are vegetation and the gray areas are urban settings. Now zoom to and view the *1978* and *2000* layers contained in the *Beijing, China* folder.

3. Based on your assessment of the Shenzhen and Beijing image sets, identify the response that is not supported.
   A. The urban areas around Shenzen are expanding by adding land along the coastline.
   B. West of Shenzhen, coastal vegetation has been replaced by an airport.
   C. There has been a large loss in agricultural lands in and around both cities in the past 30 years.
   D. The forested hills to the west of Beijing (on the far left edge of the 1979 Landsat image) have been protected from development in subsequent decades.
   E. Beijing and Shenzhen have experienced explosive growth since the late 1970s.

China's population and urban footprint are undoubtedly large and growing. However, the population of the country would be much larger if not for a government policy enacted in 1979. This policy was designed with the social, economic, and environmental problems that can be associated with a rapidly growing population in mind. The policy is controversial inside and outside of China. One unintended side-effect has been a gender disparity that has grown to epic proportions. According to the Chinese Academy of Social Sciences, China will have up to 40 million more men than women in the under 19 age group. To see a graphic illustration of this, zoom out and turn on the *sex ratio – male/female (under 15) – 2008* folder. Read the description of the statistic provided by clicking on a country's flag and then research the issue using outside resources. You could try terms such as "gendercide" or the "missing women of Asia." When you have completed this exploration, turn off the *sex ratio – male/female (under 15) – 2008* folder.

4. Where does China stand in terms of the sex ratio in 2008?

   A. It has the lowest sex ratio in the world.
   B. It has one of the five lowest sex ratios in the world.
   C. It has a sex ratio that is in the middle group of countries for the world.
   D. It has one of the five highest sex ratios in the world.
   E. It has the highest sex ratio in the world.

5. Examine the statements regarding the sex ratio and identify the statement that is *inaccurate*.

   A. Skewed sex ratios can suggest gender specific abortion.
   B. Sex ratios are more likely to be skewed in the direction of more boys being born.
   C. Sex ratio is the ratio of live fetuses to aborted fetuses.
   D. Skewed sex ratios can lead to future social problems with marriage and fertility rates.
   E. Skewed sex ratios can suggest sexual discrimination in a society.

Exploration 11.2: SHORT ESSAY

1. What factors could explain your response for question two of the multiple choice part of this exploration?

   ___
   ___
   ___

2. Explain the causes of the sex ratio of China and then describe the potential societal consequences of an unequal sex ratio.

   ___
   ___
   ___

Exploration 11.3: EAST ASIA CULTURE

East Asia stands out as one of the more culturally homogeneous regions of the world. Despite its relatively higher degree of uniformity, there remains significant cultural variation. Its similarities, however, are largely rooted in the dominant role of Chinese culture in the region. Let's begin our exploration of East Asian culture by visiting a location that helped shape the region for five centuries. Zoom to the *Forbidden City* placemark and be sure that the *3D Buildings* layer is turned on in the *Primary Database*. This complex of buildings is a UNESCO World Heritage site that was the imperial palace of China during the Ming and Qing dynasties. Survey the scene from the air and then fly into the *Inside the Forbidden City* Gigapan image.

Exploration 11.3: MULTIPLE CHOICE

1. Based on what you can ascertain from the Forbidden City placemark, World Heritage link, and Gigapan image, identify the statement that is *incorrect*.

   A. Tourists are allowed inside the Forbidden City.
   B. The Forbidden City has been recognized for its exemplary example of imperial palace architecture.
   C. The Forbidden City includes more than 10,000 rooms filled with art and furniture.
   D. The Forbidden City included defensive features such as walls and a moat.
   E. The Forbidden City is situated at an angle that is 45° off the primary axis of Beijing.

The Forbidden City and its neighbor immediately to the south, Tiananmen Square, have and continue to shape the culture of China. Click the *Tiananmen Square* placemark to view the world's largest square. A number of historic events have occurred here including the May Fourth Movement and the formal establishment of the People's Republic of China by Mao Zedong in 1949. In fact, Chairman Mao is available for viewing in the Chairman Mao Memorial Hall located in Tiananmen Square. More recently, Tiananmen Square was the site of large-scale demonstrations that resulted in the deaths of hundreds of Chinese. These demonstrations were a direct reflection of the forces of globalization at work. Research Tiananmen Square using outside resources to gain an understanding of this event.

2. Which of the following words or phrases are not directly associated with the 1989 demonstrations?

   A. government crack-down
   B. Boxer Uprising
   C. pro-democracy
   D. June 4th Incident
   E. tank man

Chairman Mao's Mausoleum gave you a taste of the cult of personality that has arisen in the region at times. As much as China can be considered the cultural foundation in the region, this is one area where North Korea has the Chinese beat. A cult of personality can be created when the leader of a country manipulates mass media to create adulation of them as a hero figure. Be sure the *3D Buildings* layer is on and open the *Pyongyang* placemark to view a classic example. This statue was erected while Kim Il-sung was in office as Premier of the Democratic People's Republic of Korea. Today, albeit deceased, Kim Il-sung maintains his position as Eternal President of the Republic. In the background you can see the Ryugyong Hotel, which has an interesting story to itself. Now open the *Arirang Festival* placemark. This placemark is centered on the Rungrado May Day Stadium. This happens to be the largest stadium in the world with a capacity exceeding 150,000. The stadium has been used to burn alive attempted assassins of the Dear Leader, as the Supreme Leader, Kim Jong-il is known. It's also used for the Arirang Festival, also known as the Mass Games. Go to YouTube and search for "Arirang Mass Games" and watch a few of the videos.

3. In the Arirang Mass Games

   A. tens of thousands of people perform various card tricks and gymnastics routines in perfect unison.
   B. hundreds of badminton games are played simultaneously.
   C. the world's largest table tennis (ping-pong) tournament takes place.
   D. only adults participate.
   E. the performance takes place solely on the fields of play.

Take a quick tour of some of the significant religious sites in the region by opening the religious sites folder and clicking each placemark. These four sites epitomize a characteristic that is often associated with eastern religious traditions.

4. Eastern religions, as supported by the imagery in the religious sites folder, often

   A. have sites that display multiple deities.
   B. favor urban environments for shrines.
   C. attempts to locate shrines in lowland areas.
   D. emphasizes large worship halls.
   E. have strong ties to the organic or natural world.

Changing the pace, let's dive into the Tokyo metropolitan area. Fly into the *Shibuya Station Crossroads* and *Shinjuku Kabuki-cho Crossing* 360° panoramas. These images give you a taste of urban life in one of the world's largest cities. As you explore the images, think of ways that this scene is different or the same from other urban settings you have viewed or experienced.

5. Upon viewing the Shibuya Station Crossroads and Shinjuku Kabuki-cho Crossing 360° panoramas, select the statement that is *not* verified by the images.
   A. There is evidence that the local writing system is ideographic.
   B. American companies have yet to become established in this marketplace.
   C. This is a culture that values and utilizes technology.
   D. There is no shortage of electricity.
   E. English is utilized for some signage.

Exploration 11.3: SHORT ESSAY

1. Research the current and the previous leader of North Korea. Why can the term "cult of personality" be applied to these men? What methods have been employed to strengthen their cult of personality?

   ____
   ____
   ____

2. What aspects of the Tokyo 360° urban panoramas stand out to you as the most different? What locations in the US or Europe can you think of that would have similar scenes?

____

____

____

Exploration 11.4: EAST ASIA GEOPOLITICS

Chinese geopolitics can be a complicated topic, particularly when one considers some of the territorial disputes between China and its neighbors. Open the GEOPOLITICS folder and then open the *territorial disputes* folder. Included in this folder are five placemarks that lead you to the general vicinity of disputed territory. Utilize Google Earth™ and outside sources to identify parties involved in the disputes and the core issues at stake for each location and then answer the following questions.

Exploration 11.4: MULTIPLE CHOICE

1. Which of the following statements regarding territorial disputes in the East Asia region is not accurate?

   A. Taiwan is claimed by the People's Republic of China and Japan.
   B. Part of the Indian state of Arunachal Pradesh is claimed by China.
   C. Aksai Chin is administered by China but is claimed by India.
   D. There is a disputed border in the region of a volcano along the North Korean and Chinese borders.
   E. Taiwan (Republic of China) claims the Spratly Islands.

2. Identify the country that does not claim part or all of the Spratly Islands.

   A. Brunei
   B. Indonesia
   C. People's Republic of China
   D. Philippines
   E. Vietnam

Now travel to the Demilitarized Zone (DMZ) along the North Korea and South Korea border. Go to the *DMZ-JSA* placemark. The DMZ is a strip of land between the two countries that acts as a buffer zone. The strip is about four kilometers wide on average. This may be the most heavily militarized border in the world as it represents the location of the war front between the North and the South at the time an armistice was signed in 1953. Because there has never been a peace treaty, the two sides are technically still at war. Since the DMZ was established there have been numerous incursions by the North into the South. The most intriguing are the numerous tunnels that have been discovered under the DMZ. Blasted through bedrock, these have been designed as invasion routes into South Korea. Above ground there are a number of curiosities as well. More than 25,000 US troops are stationed in South Korea as a relic of the Korean War and the tense border. There is an area known as the Joint Security Area (JSA) where soldiers from both sides face each other daily and any negotiations between North and South Korea take place. Study the border between North and South Korea and attempt to identify the number of routes through the DMZ.

3. At how many locations are there transportation routes (e.g., highways and/or railroads) that cross the DMZ?

A. 0
B. 1
C. 2
D. 3
E. 4

Yet another set of curious features near the DMZ are the peace villages maintained by both North Korea and South Korea. South Korea's village of Daesong-dong is a small agrarian community where villagers are exempt from taxes and military service. The North Korean village of Kijong-dong, on the other hand, is a propaganda village built in the 1950s. Its colorful buildings with electricity were supposed to look enticing to the people of the South. In reality the buildings were/are empty and the town is a farce. The two towns have had an ongoing competition in the realm of tall flag poles in recent decades. Both sides have built and extended their flagpoles to outdo the other. Go to the *flag of the North* and the *flag of the South* placemarks to view the respective sites.

4. Based on the flag of the North and flag of the South placemarks, identify the *most correct* statement.

A. The flagpole of the north is likely taller based on the size of the base of the structure.
B. The flagpole of the south is likely taller based on the size of the base of the structure.
C. The flagpole of the north is likely taller based on the length of the pole's shadow.
D. The flagpole of the south is likely taller based on the length of the pole's shadow.
E. You cannot compare them with certainty because the images were captured on different dates.

Go to the *clearly defined boundary* placemark. Be sure the *3D Buildings* layer is turned on. This boundary is one that may be second only to the DMZ as a prominent mark upon the landscape. Follow this boundary in both directions for as long as it is visible.

5. After evaluating this boundary, select the statement that is *least* correct.

A. The boundary branches in multiple directions at some locations.
B. The boundary follows some historic division that is not readily apparent on today's political landscape.
C. The boundary is greater than 1500 kilometers long.
D. The boundary is not entirely continuous along its course.
E. The boundary generally correlates with modern national and/or sub-national boundaries.

Exploration 11.4: SHORT ESSAY

1. Based upon your evaluation of the various claims to the Spratly Islands, explain which country or countries you think have the strongest claims.

____
____
____

2. Research the DMZ and zoom in to study its physical characteristics. Identify and discuss the paradoxical/ironic benefit associated with a place that is so deadly to humans.

____

____

____

Exploration 11.5: EAST ASIA ECONOMY AND DEVELOPMENT

The Great Wall of China can be considered as a cultural, geopolitical, and economic feature of the Chinese landscape. Open the ECONOMY AND DEVLEOPMENT folder and go to the *Great Wall* placemark. Zoom in and explore the cultural landscape at this location. Be sure to turn on the *Panoramio Photos* to view photographs of the area.

Exploration 11.5: MULTIPLE CHOICE

1. After viewing the cultural landscape of the Great Wall in the vicinity of Ying Beigoucun, which of the following statements is *not* accurate?

   A. The Great Wall is a source of local economic development.
   B. A toboggan slide has been constructed at the location.
   C. Multiple cable cars/chair lifts have been constructed at the location.
   D. Souvenir shops are numerous.
   E. Chinese authorities have forbidden development that is not historically relevant in the vicinity of the Great Wall.

The East Asia region has been a primary driver of the world economy for some time now. What are the prospects for this to continue in the future? One way to assess this is to look at the rate of investment on fixed assets like factories, machinery, and equipment by the economies of the region. Turn on the *investment (% of GDP) – 2008* folder and examine the statistical values for the countries of the region by clicking the individual flags. Compare these to other countries and regions of the world.

2. Based on your assessment of investment rates around the world in 2008, select the statement that is *least* accurate.

   A. East Asia is a leading region in terms of business investment in fixed assets.
   B. China has an investment rate in excess of 40% of its GDP.
   C. All the countries of the region, excluding North Korea, have investment rates higher than 20%.
   D. The US investment rate is higher than the majority of the world's countries.
   E. This suggests that East Asia's potential for economic productivity will continue an upward trajectory.

Turn off the *investment (% of GDP) – 2008* folder and turn on the *exports – 2008* folder to get a feel for the flow of goods and commodities provided to foreign consumers by domestic producers. Examine the statistical values for the countries of the region by clicking the individual flags and compare these to other countries and regions of the world.

3. Based on your assessment exports around the world in 2008, select the statement that is *least* accurate.

A. Sub-Saharan Africa lacks a single top-30 exporter.
B. Germany, China, and the US are three of the world's leading exporters.
C. The northern hemisphere countries generally have higher export figures than the southern hemisphere countries.
D. All countries in the East Asia region are among the world's ten highest exporters.
E. The top three exporting regions in the world are East Asia, North America, and Europe.

Turn off the *exports – 2008* folder and zoom to the *export 1* placemark. Explore this location and do the same for *export 2*. Review the information associated with the *export 2* link. You have seen two of the busiest ports in the world, in Hong Kong and Shanghai respectively, and gained an understanding of the primary transport and storage unit. At both these locations you see similar features as goods from across China converge on ports so they can be shipped to locations around the world. The *export 3* placemark illustrates the most essential vehicle in the transport of goods. Take a look at the linked webpage that provides information and pictures regarding these vehicles. If you continue the journey by clicking the import 1 placemark, you'll see a typical destination for these exported goods, Long Beach Port. Goods are unloaded here and placed on trucks and trains (*import 2*) where they will eventually reach distribution centers (*import 3*) and finally retail stores (*import 4*). The link associated with import 1 details the trade balance between the United States and China. Open the link to evaluate this relationship over time. At this Census Bureau Foreign Trade Statistics site, you can explore these relationships with other countries by clicking the "Country/Product Data" link at the top of the page, and then selecting "Country Index."

4. After viewing the placemarks along the export/import path and viewing the associated links, which of the following statements is *least* accurate?

A. Rail and/or highway connectivity is an essential part of the ports.
B. Each intermodal container is created in a different size and/or shape depending upon the cargo it will carry.
C. Loading and unloading of container ships takes place with multiple cranes operating on one ship.
D. The largest container ships can carry no more than 5,000 intermodal containers.
E. Container ships use a technique called containerization to carry their loads.

5. Review the trade balances between the US and its East Asia trading partners and identify the statement that is *least* accurate.

A. In the last decade, the US has steadily narrowed the trade gap with China.
B. The US has a trade deficit with South Korea.
C. The US has a trade deficit with Taiwan.
D. The US has a trade deficit with Japan.
E. The US imports about five times what it exports to China.

Exploration 11.5: SHORT ANSWER

1. Assess the local economic development and preservation of historical integrity around the Great Wall at Ying Beigoucun. Do you believe that more or less development should occur? Do you think that the aesthetic values and historical integrity of the site are adversely impacted by these developments? Explain your answer.

   ____

   ____

   ____

2. Summarize the trade patterns between the US and its East Asian trading partners. Based upon your personal knowledge and what you have gleaned from this exploration, what are the prospects for the future?

   ____

   ____

   ____

YOU MAP IT! Carbon Dioxide Emissions

Open and turn on the *YOU MAP IT!* folder. Click the Zouxian placemark to return to the coal-fired power plant we visited earlier in this exploration. Note that the location, a description of the plant as the top CO_2 producer in China, the tons of CO_2 emitted annually, and the megawatt hours of electricity produced annually have been included in the placemarks properties. Create similar placemarks for the top two CO_2 producers in Africa, Asia, Europe, North America, South America, and Oceania. Also identify and map the top two in your state or province. The information for this project is available at the Carbon Monitoring for Action website that is associated with the YOU MAP IT! folder.

Your instructor may want you to create a document with the information associated with your fourteen power plants as well as a narrative that interprets the geographic patterns of the locations you identified. For example, are the sites usually in urban or rural locations, more or less affluent regions?

Turning it in:

Your instructor will provide you with an explanation of how to submit your results from this assignment. This may include e-mailing a .kmz file you create from your *YOU MAP IT!* folder and a document with the required information to your instructor.

To create a .kmz from your *YOU MAP IT!* folder, simply click once on the *YOU MAP IT!* folder to highlight it, then go to File, Save Place As…, and save it in an appropriate location on your computer.

Name: ______________________________

Date: ______________________________

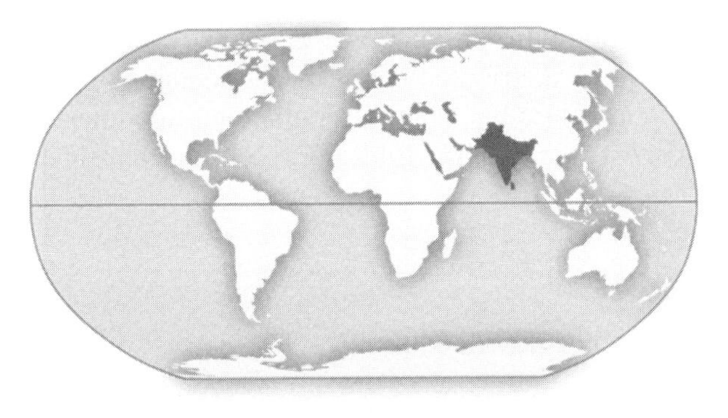

Chapter 12: South Asia

Separated from the rest of the Asian landmass by a series of high mountain ranges, South Asia consists of the countries of India, Pakistan, Bangladesh, Bhutan, Nepal, Sri Lanka, and a number of islands in the Indian Ocean.

Just as China is the dominant influence in the East Asian region, so is India in the part of the world that is often referred to as the Indian subcontinent. The population of India now exceeds one billion people and will become the most populous country in the world in the coming decades. Pakistan and Bangladesh also have very large and rapidly growing populations that currently rank sixth and seventh largest in the world, respectively. A number of the world's largest urban areas are found in the region including Mumbai, Kolkata, and Delhi in India, Karachi in Pakistan, and Dhaka in Bangladesh.

South Asia is home to a range of environmental conditions. The highest mountains of the world are found in Himalayan highlands of Nepal, while the deltas of Bangladesh are some of the most extensive lowland areas of the world. The region is home to a landmark site in the evolution of environmentalism and corporate responsibility. What happened at Bhopal India continues to shape policy worldwide.

Fertility rates in the region have declined in recent decades, but their remains enough population momentum for the region to overtake East Asia as the world's most populous in coming years. The population of South Asia, and India in particular, have also established a global presence as the Indian diaspora has spread to every corner of the world.

As South Asian populations have become established, so to have their cultural traditions. Restaurants, places of worship, and even political monuments can be seen around the world. Within the region, a number of splendid cultural features are iconic elements of the landscape. Another cultural icon is the sport of cricket. This unique game is played across the region, and around the world for that matter.

A more serious game has been played over the years between Pakistan and India as the two countries have periodically ratcheted-up tensions along the nuclear weapons front. A number of secessionist movements, territorial disputes, and civil conflicts have added to the region's classification as less than peaceful.

In spite of these tensions, there has been significant economic growth in some sectors. India has become a global leader in the development of information technology. Several cities have become hubs for Indian and multinational corporations in this field. Long-term economic growth, however, will be challenged until low literacy rates are addressed.

Download EncounterWRG_ch12_SouthAsia.kmz from **www.mygeoscienceplace.com** *and open in Google Earth™.*

Exploration 12.1: SOUTH ASIA ENVIRONMENT

In a region where the soaring peaks of the Himalayas garner much of the attention, it is the lowlands of the region that enable the population to grow with their abundant agricultural production. As the snows melt in the high mountains, they bring water and nutrients downstream to be distributed along the floodplains of the region's great rivers. Rivers such as the Indus, Ganges, and Brahmaputra have densely populated floodplains and rich deltaic environments. Turn on and expand the *Sundarban, South Asia* folder. Explore the physical landscape, paying special attention to the pattern of streams.

Exploration 12.1: MULTIPLE CHOICE

1. The Sundarban imagery illustrates a fluvial pattern that can be described as

A. headwaters.
B. a concentric drainage network.
C. a nickpoint migration zone.
D. a distributary network.
E. a collection of intermittent streams.

The Sundarban is one of the world's largest mangrove forests (the darker green area of the images). The region has very high population densities and constant threats from natural hazards. Use the *1977* and *2000* image layers to evaluate the degree to which the mangrove forest has changed over a roughly 25-year period.

2. Based on your visual assessment of the remotely sensed imagery, what percentage of the mangrove forest has been cleared between 1977 and 2000?

A. none of the forest has been cleared
B. less than 10%
C. approximately 25%
D. approximately 50%
E. more than 75%

The Sundarban region is subject to flooding threats. These can occur as a result of tropical cyclone or as a result of the spring run-off. Storm surges (the wall of water pushed in front of a tropical cyclone) can exceed 13 meters. Therefore coastal locations that are less than this height above sea level may be subject to flooding. In 1970, a storm surge in this area took the lives of more than 500,000 persons. To gain an understanding of the area that is potentially affected with these events, one must have an understanding of the topography of the land.

3. With a 13-meter storm surge, which of the following locations would *not* be directly impacted?

A. Dhaka, Bangladesh
B. Rejupara, Bangladesh
C. Jessore, Bangladesh
D. 24.27° N, 89.75° E
E. Kolkata, India

Not all hazards of the region are tied to "natural" phenomena. In fact, one of the world's worst industrial catastrophes took place in the city of Bhopal, India in 1984 when a toxic gas was released from a Union Carbide plant. More than 500,000 people were exposed and several thousand died. The incident led to dramatic advances in the realms of safety for chemical plant workers and surrounding communities. For example, the Emergency Planning and Community Right-to-Know Act of 1986 is a US federal law that deals with emergency preparedness. A major component of this legislation is the requirement to document, notify, and report potential chemical hazards that are present in communities. There are now dozens of sources where you can find out what potential hazards exist in your community. One such location is the US Environmental Protection Agency's Toxics Release Inventory Program. There are mapping and database tools at this site that will help you gain a better understanding of the chemical inventory near you. A user-friendly tool is the MyEnvironment mapper. Open the link associated with the *required reporting* placemark and enter "67449." This is a rural location in central Kansas. Explore the results page. In the upper right-hand corner, you will see a list of facility types and sites and you may be surprised to find that there are two Superfund sites.

4. One of the Superfund sites in Kansas' 67449 zip code is

A. the Tri-County Public Airport.
B. the Monsanto fertilizer plant.
C. the coal-fired power plant.
D. the abandoned nuclear facility.
E. the Army munitions depot.

The Bhopal incident led to sweeping changes in the way chemicals are managed and reported, but what does the Bhopal plant look like today? Open the Union Carbide placemark and explore the landscape and then turn on the *Panoramio* images and study the photos of the site.

5. Based on the imagery and photos available on Google Earth™, the Bhopal plant

A. has been completely cleared from its former site.
B. appears to be fully operational.
C. has been replaced by a large housing development.
D. is the site of a memorial park.
E. has been abandoned and is in a state of total disrepair.

Exploration 12.1: SHORT ESSAY

1. What natural hazard is most likely to affect the location where you live? What precautions (e.g., shelters, early warning systems) have been taken to prepare for a hazardous event?

____

____

____

2. Utilize the EPA's MyEnvironment tool to assess the environmental conditions at your location. Type your current zip code into the "Location" box and then describe some of the characteristics of your local environment, particularly features or characteristics that are cause for concern.

____

____

____

Exploration 12.2: SOUTH ASIA POPULATION

The large and growing population of South Asia is distributed unevenly across the region. South Asia's population is patterned in a fashion that helps exhibit the region's reliance on agriculture. Turn on the *population density* layer and study the patterns of density.

Exploration 12.2: MULTIPLE CHOICE

1. The largest swaths of the densely populated areas of the region most strongly correlate with

A. the high mountains.
B. the major river valleys.
C. the Deccan Plateau.
D. the Thar Desert.
E. the coastlines.

Turn off the *population density* layer. An agricultural product that is particularly important in the northern and eastern parts of India is tea. Double-click the *Agriculture in Darjeeling, India* placemark to fly into the 360° panoramic image. Darjeeling is well-known for the production of tea. Use this image and outside research to learn about the production of this commodity.

2. Which of the following statements is *least* accurate?

A. Workers involved in the tea industry are disproportionately women.
B. Darjeeling tea has been cultivated for commercial purposes for more than 2000 years.
C. Darjeeling tea is usually a variety that originally came from China.
D. India is one of the largest tea producers in the world.
E. The British East India Company played a large role in ramping up tea production in India.

Just like Indian tea, the people of India can be found in every corner of the world. The term diaspora is used to refer to a relocated group of people with some national or ethnic identity. Let's explore just a few places where Indian culture and Indian persons have made their mark on the cultural landscape. Turn on the *3-D Buildings* layer and go to the *diaspora 1* placemark. Determine where you are located and what the significance of the location is in terms of Indian heritage.

3. Select the person that best captures the significance of the diaspora 1 placemark.

A. Norah Jones
B. Ben Kingsley
C. Jhumpa Lahiri
D. Amar Bose
E. Bobby Jindal

Now go to the *diaspora 2* placemark and then open the associated link to the Statistics Canada website. Utilize the resources available from this site to evaluate the characteristics of Canada's Indian population. Clue: Immigrant population statistics will be most helpful.

4. Which of the following statements regarding the population of Indian-born persons in Canada is *incorrect*?

A. India represents the third largest source for immigrants to Canada.
B. The number of persons from India that have migrated to Canada for the periods 1991-1995, 1996-2000, and 2001-2006 steadily increased.
C. As of the 2006 Census, Canada had more than 700,000 persons that were born in South Asia.
D. More than 50,000 South Asians were found in each of the Yukon and Northwest Territories, and Nunavut.
E. Ontario had the highest population of South Asians of any Canadian province.

Now examine the other diaspora placemarks (3-6). Do some Google searches and identify why these locations have been included in our selection of Indian diaspora sites.

5. Utilizing your diaspora placemarks and supplemental research, which of the following statements is *least* accurate?

A. Many Indians came to South Africa in the 19th century to work as indentured servants.
B. People from India make up more than 15% of Alaska's population.
C. The British brought significant numbers of Indians to Fiji in the 19th century.
D. Indo-Trinidadians are the largest ethnic group of Trinidad and Tobago.
E. Close to half of the population of the United Arab Emirates is from South Asia.

Exploration 12.2: SHORT ESSAY

1. Do a Google search for "List of Indian Americans." Identify two people on the list that you feel have made particularly important contributions to the cultural, educational, political, or scientific realms in the US. Explain why you selected these individuals.

____
____
____

2. What is the most conspicuous South Asian cultural element in your community? This could be a shop or restaurant or friends you may have. How have these elements influenced your understanding of South Asia?

____
____
____

Exploration 12.3: SOUTH ASIA CULTURE

One of the defining cultural characteristics of South Asia is its Hindu religious tradition. Hinduism's hearth area is the Ganges River Valley, but shrines for this complicated polytheistic tradition can be found across the region. Go to the *Badrinath* placemark. Badrinath is a key pilgrimage site for followers of Hinduism. This place is particularly meaningful to the supreme Hindu god of Vishnu. More than half a million visitors come to this site annually. In spite of the heavy visitation, the Temple is only open for half of the year.

Exploration 12.3: MULTIPLE CHOICE

1. Why do you think this site is only open for a portion of each year?

   A. It exists in an alpine location with high snowfall.
   B. Visitors have to walk to the city.
   C. There is no bridge across the stream to the temple, so it is only accessible during low water times.
   D. There are no hotels in the city, so visitors must camp.
   E. Nepal claims the city as well, and controls it for six months of the year.

South Asia, like other parts of the world with dominant religious traditions, is also home to followers of many other faiths. At one time Buddhism was a primary religion in the region. Only in Sri Lanka does it maintain majority status. Nonetheless, important sites remain on the cultural landscape of the subcontinent. For example, go to the *Somapura Mahavira* placemark. This Bangladeshi World Heritage site was an important Buddhist monastery from the 7th to 12th century. Islam is also present in the region as the dominant religion of Pakistan and Bangladesh. There are also more than 100 million Muslims in India. Agora, India is home to one of the most recognizable features in the world, the Taj Mahal. Fly into the *Taj Mahal* 360° panorama and explore this complex that has also been recognized by the World Heritage Committee. Consult outside resources to gain a basic understanding of the site.

2. The white marble building is the central focus of the site. This building is actually

   A. one of the largest mosques in the world.
   B. a tomb for a Mughal's emperor's wife.
   C. the primary madrassa of Uttar Pradesh.
   D. a visitor center for the real Taj Mahal, the smaller red building.
   E. a British outpost from the time of colonialism.

Another religious tradition is represented at Amritsar. This is a great example of early globalization as two significant religious traditions abutted one another and were both jockeying for prominence in the region. Eventually a hybrid religion emerged that combined elements of both religions. The *Golden Temple* is the most sacred site for this religion as it rests upon the location where the founder of the religion used to live and meditate.

3. Select the answer that lists the religion that venerates the Golden Temple, and the two religious traditions that were combined to create the religion.

   A. Jainism; Islam and Buddhism
   B. Sufism; Hinduism and Islam
   C. Sikhism; Islam and Hinduism
   D. Zoroastrianism; Christianity and Islam
   E. Druze; Islam and Buddhism

Go to the *Shaheed Minar* placemark. This is a national monument in Dhaka, Bangladesh that commemorates those who were killed in a national uprising in 1952. This uprising went hand-in-hand with the changing geopolitical landscape of the era. Often times, cultural elements can spur further geopolitical change as they act as centrifugal forces. That is to say something that does not promote state unity. The events memorialized by the Shaheed Minar eventually contributed to new geopolitical realities and have remained a point of focus for Bangladeshi communities around the world. For example, go to the *Altab Ali Park* placemark and study the landscape carefully. If you look closely in the southwest corner of the park you can see a replica of the Shaheed Minar.

4. What cultural element was at the root of the conflict memorialized by the Shaheed Minar?

   A. ethnicity
   B. democracy
   C. gender
   D. religion
   E. language

The geography of sport can tell us a lot about culture. What kinds of sports are most popular and why? Are individual or team sports emphasized? Are sports that require specialized equipment or those that can be played with very limited equipment more popular? Why did a sport gain popularity in one or more regions of the world and not another? South Asia has several sports that are strongly tied to the region. One of these sports is polo. Polo is thought to have originated in the region. The first game may have occurred more than 2000 years ago. The Moguls were largely responsible for spreading the sport out of the region. A second major sport in the region gained popularity much later. Double-click the *games people play* folder. Exploring this small town, you will find a least two places this game is played. You can click the other placemarks in this folder to view some of the stadiums in the region that host this sport. One thing you will see in common among these sites is the strips of bare earth. This strip is the pitch.

5. Identify the sport and select the statement that is least accurate.

   A. Each team has eleven players.
   B. One side at a team bats, while the other bowls and fields.
   C. This sport spread in correlation with the English Empire.
   D. Points scored in the game are known as "wickets."
   E. The name of the game is cricket.

Exploration 12.3: SHORT ESSAY

1. Why do you think there is a replica of the Shaheed Minar in a suburb of Manchester, England? What evidence could you provide that backs up your hypothesis?

   ____
   ____
   ____

2. Identify a sport that is quite popular in one or more world regions but has not "caught on" in other parts of the world. Provide a statistic such as number of players or television viewership to back up your claim. Provide a possible explanation why the sport has not been as popular.

   ____
   ____
   ____

Exploration 12.4: SOUTH ASIA GEOPOLITICS

British colonial rule in India unified much of the subcontinent politically for the first time in the mid-19th century. Turn on the *India 1804* layer to view a historical map from the Rumsey Historical Map Collection. This map was made from surveys completed by the British East India Company. Use the transparency slider at the bottom of the Places pane to alter the degree to which you can see the underlying contemporary boundaries. Assess the historic map in comparison to today's political boundaries. Also note the portrayal of physical features such as islands.

Exploration 12.4: MULTIPLE CHOICE

1. Which of the following statements regarding the 1804 map of India is *least* accurate?

A. There is limited correlation between British colonies and Indian states.
B. The Himalayan Mountains are grossly exaggerated on the 1804 map.
C. The areal coverage of the Maldives on the 1804 map is too large.
D. There is a relatively strong correlation between the Indian border with Nepal on the 1804 map and the contemporary boundary.
E. The city of Hyderabad can be identified on the 1804 map.

Turn off the India 1804 map. The British colonial presence is still in the region. One curious location is the island atoll of Diego Garcia. Zoom to the Diego Garcia placemark and have a look around. This is a British territory with a fascinating 20th-century geopolitical history. Research the location utilizing outside resources and then zoom in to examine the tarmac.

2. Based upon your outside research and the imagery from Google Earth™, you can identify the numerous large planes parked on the eastern half of the tarmac as

A. American B-52s.
B. Russian Tu-22s.
C. Chinese J-10s
D. European EF-2000s
E. Indian Su-30s

The region continues to have a variety of hot spots in terms of geopolitical tensions. Some of these are internal, and some are transnational. India and Pakistan have has tense relations since their independence following the violent partition of British India in 1947. Since that time, Indo-Pakistani relations have been characterized by four wars and numerous skirmishes and ongoing territorial conflicts. The fact that both sides have nuclear weaponry makes the tense relations all the more concerning. Jammu Kashmir is one of the most disputed regions. You can view a small slice of the *Line of Control* in the region by opening the Line of Control placemark. Be sure you have the *Terrain* layer turned on as well. Then go to the *Wagah* placemark, open the associated link and watch the short video. Use outside resources to educate yourself further on these two locations and Indo-Pakistani relations in general.

3. Regarding the Indo-Pakistani border, which of the following statements is *least* correct?

   A. India has constructed a fence along the length of the border in Jammu Kashmir.
   B. Wagah is the most prominent of numerous highway and rail border crossings between India and Pakistan.
   C. At Wagah, Indian and Pakistani soldiers come face-to-face on a daily basis.
   D. The border closing at Wagah has been developed into a tourist destination.
   E. The Line of Control in Jammu Kashmir was established in 1972.

Indo-Pakistani relations contribute to the region having a lower than average state of peace. To see a quantified assessment of peace around the world, open the link associated with the *Global Peace Index* placemark.

4. Study the results of the country comparison and identify the incorrect statement.

   A. Europe has the highest state of peace.
   B. South Asia and Sub-Saharan Africa have the lowest states of peace.
   C. Canada, Japan, and Brazil all have very high states of peace.
   D. Northern South America has lower states of peace than southern South America.
   E. Bhutan has one of the highest states of peace among South Asian countries.

Now explore the values for several South Asian states by clicking the "comparison" link under the "Global Peace Index" heading. Add India, Pakistan, Sri Lanka, Bhutan, and Nepal to the "Countries to compare" box, and then click "compare countries."

5. After evaluating the comparison, select the *incorrect* statement.

   A. Bhutan has the lowest level of military capability, but also the most peaceful ranking.
   B. The potential for terrorist attacks is significant (score of 3,4,5) in all countries except Bhutan.
   C. India has the highest scores for level of functioning government.
   D. Nepal has the highest participation of women in parliament.
   E. Bhutan is the most peaceful country, but also spends the least (as a percentage of GDP) on education.

Exploration 12.4: SHORT ESSAY

1. Provide your thoughts on the border ceremony at Wagah. Do you think this is beneficial or detrimental to relations between the countries? Is this something you would be interested in seeing in person?

   ____
   ____
   ____

2. Examine the Global Peace Index rankings for the most current year. Describe some of the common features (economic, political, and geographic) of the countries that are included in the top ten (most peaceful).

   ____
   ____
   ____

Exploration 12.5: SOUTH ASIA ECONOMY AND DEVELOPMENT

South Asia is a land of economic contrasts. Great poverty coexists with rapidly rising technology and scientific sectors. The growth pole for these sectors is centered in the south of India. For example, the city of Bangalore has been referred to as the Silicon Valley or Silicon Plateau of India. Go to the *India's Silicon Valley* placemark. This centers your view on a part of Bangalore that has a number of high-tech industries.

Exploration 12.5: MULTIPLE CHOICE

1. Use the Find Businesses tab in the Search box to determine which of the following technology companies do not have a presence in Bangalore.

A. IBM
B. Dell
C. Oracle
D. Walrusoft
E. Apple

Another area of growth on the subcontinent is the automobile industry. With a rapidly emerging middle class, India has a growing demand for automobiles. The leader of Indian automobile production is Tata Motors. This company is a global player in the automobile and truck industry and continues to grow by recent acquisitions of companies such as Jaguar and Land Rover. You can see the scope of the Tata operation by viewing one of its factories at the *Tata Motors* placemark. Tata made waves in 2008 when it announced the launch of the Nano. The Nano is designed to be the least expensive automobile in the world. This small car was to be produced at a new production facility near Singur in the state of West Bengal. You can view the site by going to the *original Nano assembly plant* placemark. A new site had to be selected after a controversy forced Tata to abandon this location. The new production facility can be viewed at the *new Nano assembly plant* placemark. Open the associated link to read about the Nano from the perspective of Tata.

2. Which of the following statements about the Tata Nano is not accurate?

A. The Nano has front and side-curtain air bags.
B. The name Nano stands for superior technology and small size.
C. The Nano holds four people.
D. The Nano's top speed is 105 kilometers per hour.
E. The Nano comes in a variety of trim packages.

While South Asia is increasingly integrated into the global economy through its production of computers and automobiles, the region still relies heavily on agricultural production. Open the *labor force in agriculture (%) – 2008* folder. You can see that roughly half of the population of the region works in the agricultural sector. What are they growing? The United Nations Food and Agricultural Organization (FAO) maintain a wealth of statistics regarding food production. Open the link associated with the *labor force in agriculture (%) – 2008* folder to access one of FAO's databases. Select "India" for the country.

3. Which of the following commodities does India not rank first in global production?

A. Bananas
B. Chick peas
C. Goat milk
D. Green peas
E. Potatoes

Explore the commodity production of countries in the region by clicking on different commodities and by selecting different countries in the drop-down box.

4. Identify the *least* correct statement from the following.

   A. The top commodity produced in Bangladesh is rice.
   B. Sri Lanka is a top-five global producer of rice and plantains.
   C. Nepal is a global leader in mustard seed production.
   D. India is a top-three global leader in the production of mangoes and okra.
   E. Buffalo milk, wheat, and cow milk are Pakistan's most valuable crops in terms of total production.

South Asia will remain overly dependent on agriculture for employment until the population of the region becomes more educated. Turn off the *labor force in agriculture (%) – 2008* folder and turn on the *literacy - 2008* folder. Study this data set and then switch to the *female literacy – 2008* folder.

5. Utilize the information from the literacy folders to identify the *least* correct statement.

   A. South Asia is one of the least literate regions in the world.
   B. The Maldives has the highest literacy rate in the region.
   C. Female literacy rates are generally higher in the region than the total literacy rates (male and female).
   D. In Nepal, Bhutan, Pakistan, and Bangladesh, roughly two of every three women are illiterate.
   E. In Sri Lanka, female literacy is close to 90 percent.

Exploration 12.5: SHORT ANSWER

1. The emergence of the Tata Nano has created some controversy. Some environmentalists have voiced concern about the huge number of vehicles (and their pollution) that will be added to the roads of India as patrons move away from eco-friendly bicycles for transport. Do you think this is a valid criticism? Explain your answer and also provide a likely response from someone with the opposite opinion of yours.

   ____
   ____
   ____

2. What happened that kept the original Nano assembly plant from becoming fully operational? Research the case and provide a brief synopsis. Do you think there would have been a different outcome if this had occurred in your country? Why?

   ____
   ____
   ____

YOU MAP IT! Global Peace Index

Open the link to the Global Peace Index in the *YOU MAP IT!* folder. We are going to map the ten most and ten least peaceful countries in the world according to the most recent data available from Vision of Humanity. Create placemarks with each country's name and Global Peace Index ranking. The most and least peaceful countries in 2008 have been placemarked.

Your instructor may want you to create a document with a narrative that interprets the geographic patterns and cultural attributes of the locations you identified. For example, what relationship is there with democracy? How does the level of development relate to the patterns? In ten years, how will these lists have changed?

Turning it in:

Your instructor will provide you with an explanation of how to submit your results from this assignment. This may include e-mailing a .kmz file you create from your *YOU MAP IT!* folder and a document with the required information to your instructor.

To create a .kmz from your *YOU MAP IT!* folder, simply click once on the *YOU MAP IT!* folder to highlight it, then go to File, Save Place As…, and save it in an appropriate location on your computer.

Name: ______________________________

Date: ______________________________

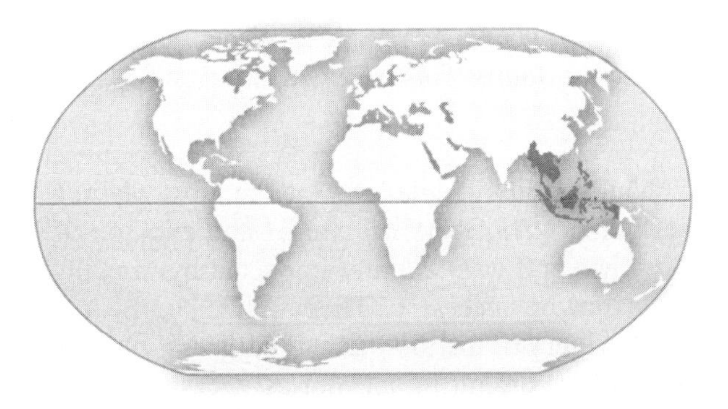

Chapter 13: Southeast Asia

Southeast Asia consists of countries that have significant social, cultural, and economic variation. Southeast Asia can be defined into two subregions: the mainland region and the insular region. The mainland region consists of Thailand, Laos, Myanmar, Vietnam, and Cambodia while the insular or island part of the region includes Malaysia, Indonesia, the Philippines, East Timor, and the microstates of Singapore and Brunei. Vietnam is the most populous country and Ho Chi Minh City and Bangkok are the largest cities.

The environment of the region is characterized by high relief and tropical climates. The region is prone to volcanoes and earthquakes and their related hazards. The moist environment produces resources such as tropical hardwoods that are continually exploited.

Agriculture in the form of rice cultivation has sustained people of the region for millennia. These unique cultural landscapes represent intensive agriculture. The pressures on the landscape are great across the region. The region has a history of environmental collapse per mismanagement of natural resources by local populations. The remnants of the great cultures can be seen in places such as Angkor Wat in Cambodia.

Modern cultures are influenced by this rich history that contains significant influence from both South Asia and East Asia along with more recent colonial influences from Europe. This cultural amalgam makes the region a popular tourist destination. Beyond the natural and cultural wonders of the region, some people come to engage in sex tourism. Most notorious in the Thai city of Bangkok, sex tourism is often tied to a global problem that is on the increase, human trafficking.

Although the Cambodian-Thai border is porous in many areas, there have been recent tensions between the two neighbors tied to ownership of an important cultural site. Both countries have promoted their perspectives on the issue on their national web portals. A survey of the web-presence of countries in the region can provide some interesting insights into government values as well as the level of development.

The levels of development in the region generally progressed rapidly in the 1980s and 1990s, particularly in Thailand, Malaysia, and Indonesia. Since the late 1990s, the region's economic progress has been uneven. Exports to the world market still fuel the economies of the region. One economy of note is Singapore. Singapore is a very small state of only 619 square kilometers. In spite of its small size, it has grown by producing efficient high-tech and service industries to pair with its role as a global export hub. Singapore is a world model for overcoming challenges to development.

Download EncounterWRG_ch13_EastAsia.kmz from ***www.mygeoscienceplace.com*** *and open in Google Earth™.*

Exploration 13.1: SOUTHEAST ASIA ENVIRONMENT

The natural hazards of Southeast Asia are an inherent part of life in the region. The 2004 Indian Ocean Tsunami was a natural disaster that was a direct result of the geologic activity in the region. The earthquake had a magnitude of approximately 9.2 and is one of the most powerful quakes ever recorded. The quake was also exceptionally long, as the earth shook for nearly ten minutes. When the quake occurred significant portions of the sea floor were displaced upward along a major fault. This resulted in a subsequent displacement of water that triggered the resultant tsunami.

The Indian Ocean Tsunami is estimated to have killed more than 200,000 people. One of the most severely affected areas was the province of Aceh, Indonesia. Open the 13.1 ENVIRONMENT folder and click on the *Banda Aceh* placemark. Banda Aceh is the capital of Aceh province and was severely impacted by the 2004 earthquake and tsunami. Study the landscape, zooming in and out as necessary. Then utilize the historic imagery capabilities of Google Earth™ by turning on historical imagery in the toolbar and adjusting the time slider to the January 27, 2005 imagery. Study the landscape again, zooming in and out as necessary. Now move the slider back to June 22, 2004 to view the city prior to the tsunami.

Exploration 13.1: MULTIPLE CHOICE

1. Which of the following statements is not supported by your study of Banda Aceh?

   A. The coastline of Banda Aceh was permanently altered from the tsunami.
   B. Much of the town has been rebuilt.
   C. In the rebuilding, there is no evidence of improved infrastructure like road upgrades.
   D. Vegetation removed by the tsunami has yet to fully recover.
   E. People still live in homes that are only a few meters above sea level.

Open the link associated with the *Banda Aceh* placemark. An image from the National Oceanic and Atmospheric Association (NOAA) illustrates the travel time for the tsunami to reach different locations around the region and world. You may be surprised to know that a tsunami can travel all the way across an ocean basin and have an effect on a far-off place. You might also be surprised to see how quickly that tsunami can travel.

2. Using the NOAA image as a guide and Google Earth™ to determine approximate distance, how fast was the tsunami moving between its point of origin as indicated by the stars on the NOAA image and Xaafuun, Somalia?

   A. 25 kilometers an hour
   B. 125 kilometers an hour
   C. 340 kilometers an hour
   D. 650 kilometers an hour
   E. 1,100 kilometers an hour

Now go to the Sri Lanka tsunami placemark and utilize the same technique of image analysis for pre and post tsunami landscapes.

3. Which of the following statements is not supported by your study of historical imagery of the Sri Lanka tsunami location?

   A. Sri Lanka did not experience significant tsunami-related destruction because of its greater distance from the earthquake.
   B. Vegetation in Sri Lanka seems to have recovered more dramatically than in Banda Aceh.
   C. Temporary shelters are evident in the 2005 imagery.
   D. A canal that connects a nearby lake (Svan Lake) and the Indian Ocean appears to have been impacted by the tsunami.
   E. A protected port has been added to the Sri Lankan location since the tsunami.

Open the link associated with the Sri Lanka tsunami placemark to view a NOAA image of wave amplitude (wave height) associated with the tsunami.

4. What region did not experience coastal wave amplitudes of at least 0.5 meters?

   A. North America
   B. Latin America
   C. East Asia
   D. Sub-Saharan Africa
   E. North Africa/Southwest Asia

Zoom to, and then open the *Papua, Indonesia* folder. You will be looking at a LANDSAT image of one of the least densely settled islands in the region. However, population and impacts on the environment are growing. One area of tremendous growth in recent years has been the development of palm oil plantations. Indonesia and Malaysia are the leaders in the production of palm oil. Palm oil has uses in food, but is increasingly seen as a source of biofuel. While biofuel can provide energy with reduced greenhouse gas emissions, it is important to understand that often times, forests are cut down to make way for the production of a biofuel source like this palm oil, or sugar cane, or corn. Turn on the 2000 and 2002 images to view the progression of this palm oil plantation.

5. Based on the *Papua, Indonesia* image sequence, which of the following statements is not supported?

   A. While the area under cultivation has increased, the road network has not grown in size.
   B. There are only minimal signs of development in the area in the 1990 image.
   C. Roughly the same amount of expansion occurs between 1990-2000 and 2000-2002.
   D. The plantation is laid out in a grid network.
   E. The plantation has expanded to the edge of a significant stream by 2002.

Exploration 13.1: SHORT ESSAY

1. If a tsunami formed in Indonesia can reach Africa in only six hours, what kinds of problems might that create for people in a very underdeveloped country like Somalia? For example, why would these people be more at risk for death or injury from the actual event and the after effects?

   ____
   ____
   ____

2. Tsunamis can present risks to the entire globe from an emanation point. What have the governments of the world done to mitigate these risks in the wake of the 2004 tsunami?

____

____

____

Exploration 2: SOUTHEAST ASIA POPULATION

The population of the region is growing more quickly in the insular part of the region and Indonesia has grown to the point of being the world's fourth most populous country. Across the region, people have relied on subsistence agriculture for millennia. A staple crop has always been rice. The production of this commodity has historically centered on the lowlands. Play the *SE Asia rice tour* to view an area of rice production in the Philippines that was established more than 2000 years ago. Go back and explore the landscape further upon completion of the tour.

Exploration 13.2: MULTIPLE CHOICE

1. The *SE Asia rice tour* suggests

   A. that rice is only grown in the flatland areas along streams.
   B. that rice is grown in sufficient quantities to exceed subsistence levels.
   C. that rice is also grown in highland areas.
   D. that the construction of rice terraces would require enormous labor.
   E. that some hillsides may contain more than 40 discrete terraces.

Southeast Asia has been able to support complex civilizations because of its historic ability to produce a surplus of food. One such example is the Khmer culture of modern-day Cambodia. The Khmer civilization reached its highpoint during the 11th and early 12th centuries. At this time the temple complex at Angkor was constructed. The site has both Hindu and Buddhist tradition associated with it. After the complex was completed, the Khmer civilization declined as a result of invasion by its Thai neighbors and overexploitation of environmental resources. Go to the *Angkor* folder and explore the site. Use the *3-D Buildings* layer to provide a realistic portrayal of the main temple, Angkor Wat.

2. Which of the following statements is *least* accurate?

   A. Local agriculture continues to take place in the immediate vicinity of Angkor.
   B. Contemporary imagery suggests that there was likely much less surface water available at the site in past times.
   C. Deforestation has occurred in the vicinity of Angkor.
   D. Multiple temples are part of the Angkor site.
   E. The architecture at Angkor Wat is distinct.

While Angkor is no longer an imperial hub, it remains essential to the Cambodian people for its role as a tourist destination. Now fly into the 360° views inside the Angkor folder (*360 A*, *360 B* and *360 C*). Assess the cultural landscape within the site.

3. Which of the following statements is not supported by your study of Angkor?

A. Angkor is a popular tourist destination for Asian visitors.
B. Local people dressed in traditional attire can be seen at Angkor.
C. If you visit Angkor, no cameras or video cameras are allowed.
D. The sun is strong, so a hat or umbrella is a good idea to bring to Angkor.
E. There is some evidence of restoration and/or reconstruction taking place at the site.

Angkor is but one of many sites around the world that were once the centers of great empires but later collapsed with environmental change and/or mismanagement on the part of humans as part of the story. Explore the sites of Tulum, Rapa Nui, Timbuktu, and Chaco Canyon. Examine the 360° views and associated links. You can turn on the *Panoramio Photo* layer to see images in and around these sites. Also use the web to do any outside research that is necessary to answer the following questions.

4. Which of the sites has the most advantageous geographic position as a hub for trade?

A. Angkor
B. Chaco Canyon
C. Rapa Nui
D. Timbuktu
E. Tulum

5. Which of the site's collapse is most clearly tied to nearly total deforestation?

A. Angkor
B. Chaco Canyon
C. Rapa Nui
D. Timbuktu
E. Tulum

Exploration 13.2: SHORT ESSAY

1. Research the construction and maintenance of rice terraces in Southeast Asia. Compare and contrast this type of agriculture and its associated labor and capital needs with what is typical in the Great Plains of the US and Canada.

___

___

___

2. Of the four sites (Angkor, Chaco Canyon, Rapa Nui, and Tulum) which environment would seem to have the most natural resources readily available to sustain a local population? Explain your answer.

___

___

___

Exploration 13.3: SOUTHEAST ASIA CULTURE

Southeast Asia's cultural diversity stems from the fact that it has been a region where cultural influences from around the globe have come together and mixed. One of the more dominant cultural threads of the region, however, is Theravada Buddhism. This school of religious thought originated from Sri Lanka but has spread over much of mainland Southeast Asia. Fly into the *Big Buddha* 360° panorama and survey the site. Exit the photo and survey the surrounding scene. You can use the *Panoramio Photos* layer or any additional *360° Cities* panoramas to determine what is interesting about the space that this religious monument occupies.

Exploration 13.3: MULTIPLE CHOICE

1. The Big Buddha is located

   A. next to a large mosque.
   B. within the confines of a minimum security prison.
   C. in front of a shopping mall.
   D. in the courtyard of a Catholic cathedral.
   E. on the grounds of city hall.

Turn on the 3D Buildings layer and go to the *Jakarta* placemark. We have left mainland Southeast Asia and flown to the modern city of Jakarta, Indonesia.

2. If you were a typical inhabitant of Jakarta what type of place of worship would you be most likely to attend? Remember you can use the *Find Businesses* search to get a rough idea of how many houses of a particular type of worship are found in a given place.

   A. a Catholic cathedral
   B. a Jewish synagogue
   C. a Muslim mosque
   D. a Hindu temple
   E. a Protestant church

Let's change the topic and fly into the *Bangkok entertainment* 360° panorama. Bangkok is known for its lively night life. Study this curious club scene and utilize any outside resources to answer the following question.

3. The Bangkok Club panorama alludes to something that is strongly associated with Bangkok culture, particularly districts like Patpong and Nana Plaza.

   A. Food is always to be eaten without shoes.
   B. Red Light Districts are plentiful in Bangkok.
   C. Many hotels in Bangkok are communal.
   D. In DJ Districts, DJs are found in most establishments.
   E. Christian coffeehouses are quintessential Bangkok.

Several locations in Southeast Asia have become hot spots for something known as sex tourism. This is the activity of traveling to specific locations to engage in sexual activity with prostitutes. People may engage in this activity because it is legal in the destination country, it may be inexpensive, or they may seek access to child prostitution. Within this region, Bali, Indonesia and Phuket and Bangkok, Thailand are associated most strongly with this activity. While there are numerous concerns associated with this growing activity, the contribution to human trafficking is a primary one. Human trafficking occurs when people are acquired by improper means such as force, fraud, or deception and then removed from their country of residence and forced to work for little or no payment under exploitative terms. Open the link associated with the *human trafficking hot spot* and read the information from the United Nations Office on Drugs and Crime.

4. The UN definition of human trafficking defines the purpose of the activity to include any of the following except

A. forced labor.
B. genocide.
C. prostitution of others.
D. the removal of organs.
E. sexual exploitation.

Now open and explore the link associated with *human trafficking in the US*. Return to the human trafficking in the US placemark and turn on the *Street View* layer in the *Primary Database*. Double-click one of the camera icons on Bellevue Avenue to see the street view. Work your way down the street by double-clicking successive camera icons. Manipulate your view to see what the businesses and landscape looks like. You are viewing an area that has been associated with human trafficking.

5. Based upon your analysis of the cultural landscape and any outside resources, which of the following scenarios in human trafficking is most likely associated with this section of Bellevue Avenue?

A. Trafficked individuals forced to work as farm laborers.
B. Trafficked children forced to work in a sex club.
C. Trafficked individuals' organs harvested in a local clinic.
D. Trafficked individuals forced to work in exploitative prostitution.
E. Trafficked individuals of certain ethnicity murdered and buried in a mass grave.

Exploration 13.3: SHORT ESSAY

1. Provide insight into the dominant religious traditions in the Philippines and Indonesia. What historical events occurred that influenced the respective religious trajectories of these locations?

____
____
____

2. Based on information you can glean from the Polaris Project, to what extent is human trafficking occurring in the United States? Describe some of the scenarios where this activity has been documented domestically?

____
____
____

Exploration 13.4: SOUTHEAST ASIA GEOPOLITICS

Transitioning to geopolitics with a cultural undertone, open the GEOPOLITICS folder and go to the *Cambodian-Thai border dispute* placemark. Explore the layers contained in the Primary Database to identify the significance of this disputed site.

Exploration 13.4: MULTIPLE CHOICE

1. Which of the following statements regarding the disputed site is inaccurate?

A. The site is a Khmer Temple.
B. The site has been designated a World Heritage Site by the United Nations.
C. During different periods the site has been part of Cambodia and Thailand.
D. Cambodian and Thai soldiers have been killed at the site from 2008-2010.
E. The site is only accessible from the Cambodian side of the border.

For the remainder of this look at geopolitics, we will examine the presence of Southeast Asia's governments on the web. Links to the English version of ten government's web portals are provided for you in Google Earth™. You are encouraged to view the websites in their official languages, as well. There is usually a small link in the upper corner of home pages that allow for this switch.

2. Which of the following countries refers to their leader as "His Majesty, the Sultan"?

A. Brunei
B. Indonesia
C. Malaysia
D. Philippines
E. Singapore

3. In your assessment of the government websites, which of the following countries is the most active in promoting tourism on its home page?

A. Brunei
B. Cambodia
C. Indonesia
D. Malaysia
E. Philippines

4. One Southeast Asian country stands out among its peers for its lack of freedom under its military-led government. Perusing the web pages of the region, it is evident that easy access to information and attractive presentation of government functions is not a priority. This country is

A. Brunei.
B. Indonesia.
C. Myanmar.
D. Singapore.
E. Thailand.

You can find all kinds of interesting information when you take a few minutes to dig into these websites. For example, go to the Vietnam portal and click the "About Vietnam" link. Then open the "national flag, emblem, anthem, declaration of independence" link and read the Declaration of Independence. You can click the "Click here to get online" link to listen to President Ho Chi Minh read the document.

5. After reading the Vietnamese Declaration of Independence, select the *incorrect* statement.

   A. President Ho Chi Minh read the Declaration of Independence in 1945.
   B. The document praises the French throughout.
   C. The opening statement of the Declaration of Independence is drawn from the US Declaration of Independence.
   D. The document draws from the Declaration of the French Revolution.
   E. The document condemns the Japanese throughout.

Exploration 13.4: SHORT ESSAY

1. After studying the websites of the governments of Southeast Asia, describe the most esoteric or unique element you discovered that is different from your home culture. This could be anything from a strange collection of images to political ideology. Explain why you would not commonly expect to see, hear, or read this element where you live.

   ____
   ____
   ____

2. Which one of the government portals do you think is most effective? Consider aesthetics, links to additional resources, and any additional criteria you deem relevant.

   ____
   ____
   ____

Exploration 13.5: SOUTHEAST ASIA ECONOMY AND DEVELOPMENT

Singapore became a sovereign state in 1965 when it seceded from newly independent Malaysia. Singapore has built upon several key strengths; it has a multiethnic, well-educated population and an exceptionally good location for trade. On the other hand, it has had to overcome some significant obstacles. Its tropical climate led to problems with malaria, and most significantly, it has very limited space and natural resources. Therefore Singapore has utilized its location to become one of the world's busiest ports as it serves trade moving from Southeast and East Asia toward destinations around the globe. Go to the *location, location, location* placemark in the ECONOMY AND DEVELOPMENT folder. Identify this location and research the site further using outside resources.

Exploration 13.5: MULTIPLE CHOICE

1. Which of the following statements about the location, location, location placemark is incorrect?

   A. This is a narrow stretch of water between Malaysia and the Indonesian island of Sumatra.
   B. This could be referred to as a global choke point.
   C. This is the main shipping channel between the Indian Ocean and the Atlantic Ocean.
   D. Piracy is still a problem in this region in the 21st century.
   E. At one point, the strait narrows to less than three kilometers wide.

There are four placemarks (*resource A-D*) that identify resources key to the economic success of Singapore. These resources are either processed in Singapore or shipped through Singapore on their way to any number of destinations around the world. Additionally, there are four placemarks (*processing and production 1-4*) that represent locations that are a stop-over point or final destination for goods that have been processed or shipped via Singapore. Examine these sites and determine their relationships. A few helper terms include IKEA, McCormick and Bridgestone.

2. Identify the correct pairings of resources with their processing and/or production sites.

   A. A-3, B-1, C-4, D-2
   B. A-1, B-2, C-3, D-4
   C. A-2, B-3, C-4, D-1
   D. A-4, B-1, C-2, D-3
   E. A-2, B-1, C-3, D-4

3. What resource is significant to resource D and its associated processing and production site?

   A. Petroleum
   B. Spices
   C. Tropical birds
   D. Rubber
   E. Timber

Another reason Singapore has been so successful economically is that they have demonstrated effective governance. One way this can be measured is the degree of corruption found in government institutions. An organization known as Transparency International works to end corruption. One of the ways they do this is providing highly respected annual reports on the state of corruption around the globe. Open the *Transparency International* link. You are encouraged to explore the site but eventually you will need to open the link to the corruption perceptions index (CPI) located on the left-hand side of the home page. The video on the page provides a great summary of the state of corruption for the current year. Click the map and an interactive version will open in a new window. As you mouse over individual countries you will see a score indicating the public's perception of corruption within each country. Higher scores suggest less corruption while lower scores suggest more. You can zoom in to the map by clicking on it.

4. What Southeast Asian country has the highest CPI (least corruption) score?

   A. Brunei
   B. Laos
   C. Myanmar
   D. Singapore
   E. Vietnam

5. Which of the following statements is not supported by the CPI map?

   A. North America and Northern Europe are two of the least corrupt regions.
   B. Central Asia is one of the most corrupt regions.
   C. New Zealand is one of the least corrupt countries.
   D. Afghanistan is one of the most corrupt countries.
   E. In general, the largest countries are the least corrupt.

Exploration 13.5: SHORT ANSWER

1. What has Singapore done to deal with corruption? Describe the current state of corruption in Singapore compared to the rest of the region and hypothesize how this has related to the economic fate of the state.

   ____

   ____

   ____

2. Provide a brief description of what the term corruption entails and then describe a situation where you have witnessed this. This could be a personal experience, a local political situation, or something you read about in the newspaper.

   ____

   ____

   ____

YOU MAP IT! Supply and Production Connections

In the *YOU MAP IT!* portion of this exploration, you are to identify another location that has historically operated as a global entrepôt, not unlike Singapore. You will fill the *YOU MAP IT!* folder with nine placemarks. The first placemark will identify the location you have chosen. Using the activity you completed in the ECONOMY AND DEVELOPMENT folder of this exploration, you will identify and placemark four resource points and four related processing and production points. Be sure to give the placemarks appropriate names and differentiate the two groups with different placemark icons.

When you complete the Google Earth™ portion of this assignment, create a document that briefly explains the relationships between each pair of placemarks to turn-in with your .kmz file.

Turning it in:

Your instructor will provide you with an explanation of how to submit your results from this assignment. This may include e-mailing a .kmz file you create from your *YOU MAP IT!* folder and a document with the required information to your instructor.

To create a .kmz from your *YOU MAP IT!* folder, simply click once on the *YOU MAP IT!* folder to highlight it, then go to File, Save Place As…, and save it in an appropriate location on your computer.

Name: ______________________________

Date: ______________________________

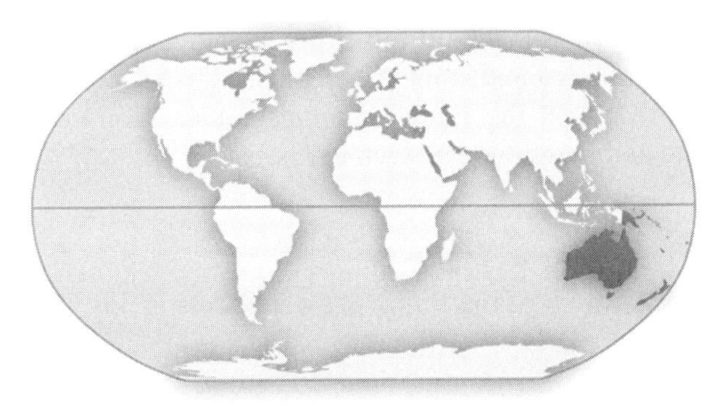

Chapter 14: Australia & Oceania

Dominated by the waters of the western Pacific, Australia and Oceania is a vast and culturally complex region. Australia dominates the region in terms of land mass as well as in economic and political realms. Beyond the islands of New Zealand, the remainder of the region is made up of small and numerous islands separated by the waters of the Pacific Ocean.

The region has challenges associated with its role as part of the Pacific Ring of Fire. Volcanoes, earthquakes, and tsunamis are all threats to inhabitants. Additionally, the region is dealing with one of the tangible effects of climate change, the rising levels of the sea. This is particularly problematic to the low islands of the region that often fail to exceed a few dozen meters in elevation. An additional concern has been the increasing incidence of wildfire associated with drought in Australia.

The population of the region is vulnerable to sea-level rise due to its close proximity to the ocean, but also to degradation of ocean resources through pollution. While many island populations in the region have capitalized on the postcard settings that are abundant, not all islands have the scenery and infrastructure to support a tourist-based economy.

Some islands are very crowded with strained resource bases. On the other hand, there are some areas in Australia's Outback that have exceptionally low population densities. This combined with the great distances between population centers dictates some unique approaches to transportation.

Other idiosyncrasies of the region include a culture that is wholly intertwined with a water-based sport. Some people would even call it a religion with a host of sacred sites scattered about Australia and Oceania. Sacred places for the indigenous peoples of the region can also be visited. These pay testament to the region's rich cultural traditions. Aboriginal, Maori, and a wealth of Polynesian and Melanesian cultures have mixed with colonial influences since the 19th century.

The colonial impacts on the region have manifested themselves in ways as mundane as the sides of the road locals drive on and as insidious as the cancers that plague Pacific Islanders as a result of extensive nuclear testing. While the days of testing have passed, controversy has not left the region. There remain calls for independence from some political entities in the region.

For example, some locations believe they are disproportionately funding their respective governments by the development of local resources. Economic resources are not solely tourism-based. The region has significant mineral and fishing resources.

Download EncounterWRG_ch14_AustraliaOceania.kmz from ***www.mygeoscienceplace.com*** *and open in Google Earth™.*

Exploration 14.1: AUSTRALIA & OCEANIA ENVIRONMENT

The Pacific Ring of Fire is a peripheral zone along the outer margins of the Pacific Ocean Basin that is prone to earthquakes and volcanoes. This tectonic activity can also spawn tsunamis. From Mount Saint Helens in Washington to Mount Fuji in Japan and Mount Ruapehu in New Zealand, this is the most active tectonic region in the world. Expand the ENVIRONMENT folder and turn on the *Plate Tectonics* folder. You can click on the individual yellow circles to get a report on the date and magnitude of each earthquake.

Exploration 14.1: MULTIPLE CHOICE

1. Based on your assessment of the Plate Tectonics folder, which of the following statements is *not* accurate?

A. The Hawaiian Islands are a result of a convergent plate boundary.
B. Hot spots have been identified on Australia, Samoa, and Tahiti.
C. Both convergent and divergent boundaries are found in the Australia and Oceania region.
D. The North Island of New Zealand is more volcanic than the South Island.
E. There have not been any volcanic eruptions on Kiribati since 1900.

2. If you were a person who was terrified of earthquakes and volcanoes, which location would be the best for you?

A. Victoria State, Australia
B. Brisbane, Australia
C. the capital city of New Zealand
D. Vanuatu
E. Tonga

Turn off the *Plate Tectonics* folder and open the *Wyperfeld National Park – Australia* folder. This park is made up of dry scrubland, what's known as Australian bush. The landscape has been managed for thousands of years by Aboriginal people. Study the northern halves of the 1973 and 2004 images and complete any outside research necessary to determine the agent that creates such significant marks on the landscape.

3. The principal agent that has created the distinctive marks on the landscapes of Wyperfeld National Park is

A. center-pivot irrigation.
B. livestock grazing rotation.
C. fire.
D. managed flooding.
E. micro-siviculture.

Sydney is Australia's largest city, but has significant constraints to the continued expansion of its urban area. Open the *Sydney, Australia* folder and view the *1975* and *2002* images of the Sydney metropolitan area. Also view the area without the dated images and develop an understanding of where the city is growing and why.

4. Which of the following would not be considered a constraint to the spatial expansion of the Sydney metropolitan area?

   A. North America
   B. Blue Mountains to the far west.
   C. Deep water inlets to the north and south of the city center.
   D. Pacific Ocean to the west.
   E. Lane Cove National Park to the northwest.

As the city grows closer to the Blue Mountains, a hazard associated with the Australian bush becomes a greater risk, bush fires. Australia has been and always will be prone to periodic, recurrent drought that can exacerbate the fire season. In recent years, some suburbs of Sydney have been severely damaged or destroyed by seasonal fires. Utilize outside resources and your knowledge of global climate regions and seasons to determine the highest risk season for fires in Sydney.

5. Evaluate the following statements regarding bush fires in Australia to identify the *incorrect* statement.

   A. Fires are most likely to occur during the summer.
   B. The 2009 bushfires in Victoria left more than 200 dead.
   C. Sydney has a mild, mid-latitude climate.
   D. The worst of the bush fires are most likely to occur in December, January and February.
   E. The subtropical desert climate immediately north of Sydney exacerbates the fire season.

Exploration 14.1: SHORT ESSAY

1. Explain why Australia has such a dearth of earthquakes and volcanoes. Are there any other areas of comparable size with such limited earthquake and volcanic activity? What is the common thread between these locations?

   ____
   ____
   ____

2. What strategies can be employed by governments and by individual homeowners to reduce the risk of losses from wildfire?

   ____
   ____
   ____

Exploration 14.2: AUSTRALIA & OCEANIA POPULATION

The people of Australia and Oceania stand to be impacted more directly than any other region by rising sea levels. The region's low islands, found disproportionately in Micronesia, will be impacted the most. The risk of flooding hazards is increasing not only as a result of rising seas, but also as a result of coral bleaching that is occurring as a result of warming seas. The low islands are coral in origin and often rely on offshore coral to buffet the effects of storms. When coral is bleached, it dies and later erodes, thus eliminating the protective ring around many of the region's low islands. Open the POPULATION folder and then open the *Pacific islands* folder. Evaluate the physical locations of the four placemarks contained in the folder.

Exploration 14.2: MULTIPLE CHOICE

1. Based on your evaluation of the Pacific islands placemarks, which of the following statements is *least* correct?

A. Rongelap and Wallis and Futuna are coral atolls.
B. Guam is not an atoll, but does have a visible reef structure around much of its periphery.
C. Most of Rongelap is less than 5 meters above sea-level.
D. The citizens of Guam are most at risk from sea-level rise.
E. The people of Tuvalu must travel more than 1,000 kilometers to reach a location with a sea-level exceeding 100 meters.

While the region of Australia and Oceania may be known for its idyllic beaches and tourist hot spots, the population of the region often lives in conditions that are the antithesis of a dream vacation. Examine the Port Moresby Housing placemark and then fly into the *Marshall Islands 1* and *Marshall Islands 2* 360° panoramas.

2. What statement best characterizes the Port Moresby housing?

A. High-rise luxury condominiums.
B. House boats.
C. Informal housing with limited infrastructure.
D. Master-planned community.
E. Low density, green design.

3. Which of the following statements does not describe a potential hazard for the residential areas visible in the Marshall Islands 1 and 2 panoramas?

A. Tsunami due to immediate adjacency to ocean.
B. Sea-level rise due to very limited gain in elevation from sea to structures.
C. Biological hazards from human decay as a result of cemetery eroding into sea.
D. Pollution from excessive build up of trash/waste on shoreline.
E. Landslides from cliffs that abut the shoreline.

Nauru is a very small independent state along the Equator in Micronesia. The state has a fascinating economic history of boom and bust. The people of Nauru face a host of challenges in the 21st century. Open the link to the CIA Factbook associated with the *Nauru's unique circumstances* placemark and then study the landscape in Google Earth™ to educate yourself about this distinctive corner of our globe.

4. Which of the following statements regarding Nauru is least accurate?

A. Nauru's economy was historically based on the exploitation of phosphate.
B. The area of Nauru is only 21 square kilometers.
C. Fresh water is plentiful in this tropical location.
D. The central part of the island is largely a wasteland from phosphate mining.
E. The population of Nauru is less than 25,000.

One part of the region that stands out for its low population density is Australia's Outback. The Outback refers to the arid interior of much of Australia. In the marginally moister zones of the Outback, sheep and cattle are grazed on large ranches or "stations." There are few roads in the Outback and goods that must be moved from one side of the continent to the other attempt to maximize efficiency. Take a look at the *curious transport 1* and *curious transport 2* placemarks.

5. What term is associated with the features that are highlighted in the curious transport 1 and curious transport 2 placemarks?

   A. Aussie snake
   B. Snowman
   C. Panamax
   D. Airbus A380
   E. road train

Exploration 14.2: SHORT ESSAY

1. Examine the detritus on the Marshall Islands beaches and identify at least four types of trash. Why do you think trash accumulates on these beaches? Why don't the tourist beaches of the region have similar problems?

   ___
   ___
   ___

2. Provide a few details about the transport devices illustrated in the curious transportation placemarks. Why wouldn't you see these moving down a highway in the United States or Europe? What makes these acceptable in the Outback?

   ___
   ___
   ___

Exploration 14.3: AUSTRALIA & OCEANIA CULTURE

Australia and Oceania represent a cultural crossroads. It is a huge region where groups of people have migrated in and out and intermingled for tens of thousands of years. There are great dissimilarities and also significant areas of common ground. One cultural tradition that can be found globally, but is associated most strongly with this region is highlighted by the places contained in the *Epic Locales* folder within the *Culture* folder. Be sure your *Terrain* layer in the *Primary Database* is turned on and then go to the four locations highlighted. You can search for these place names on the web to gain more information about their significance.

Exploration 14.3: MULTIPLE CHOICE

1. Identify the term or phrase that does *not* fit with the cultural tradition associated with the *Epic locales* placemarks.

   A. men in grey suits
   B. terminal velocity
   C. goofy foot
   D. riptides
   E. hang ten

While certain cultural activities can be found virtually anywhere around the world, certain activities are aided and abetted by the configuration of the physical environment. In fact, all of the sites in the Epic Locales folder have configurations that enhance the features that the members of the previously discussed culture value.

2. The sites in the Epic locales folder are known for

A. big wave surfing.
B. snorkeling.
C. deep-sea diving.
D. swimming with dolphins.
E. whale breaching.

Open the *Sacred Places* folder and view three important locations for indigenous groups of the region. Utilize your texts and any outside resources to determine the traditional culture that is associated with each site.

3. Identify the correct pairings of traditional cultures and sacred places from the following choices.

A. Cape Reinga – High Islanders; Halema'uma'u Crater – Fijians; Uluru - Kanakas
B. Cape Reinga - Micronesians; Halema'uma'u Crater – Samoans; Uluru - Polynesians
C. Cape Reinga – Low Islanders; Halema'uma'u Crater – Tongans; Uluru - Samoans
D. Cape Reinga - Maori; Halema'uma'u Crater – Hawaiians; Uluru - Aborigines
E. Cape Reinga - Kanakas; Halema'uma'u Crater – Polynesians; Uluru - Fijians

Traditional and Europeanized groups in the region both find great value, whether it is symbolically or economically, in the vast resources of the sea. A number of governments, organizations, and individuals are collecting data and information about the oceans and increasingly making it available in readily accessible formats in Google Earth™. For example, you can view the path of a striped marlin that was tagged with a sensor that collects information on sea surface temperatures, depth, and location. This data is transmitted to satellites when the animal breaks the surface of the ocean. This data can be grouped with other tagged animals to locate migration corridors, feeding areas, and breeding grounds for respective species. This can help governments better manage the resources of the sea. Double-click the *fish-tracking* path in the CULTURE folder and assess the path that this striped marlin followed.

4. Which of the following statements regarding the path of the striped marlin that is illustrated with the fish tracking path is *least* accurate?

A. The marlin was tracked for more than 5,000 kilometers.
B. At one point the marlin reversed its course and swam in the opposite direction of its primary movement.
C. The marlin's track ends near New Zealand's North Island.
D. The marlin crossed the Tropic of Cancer along its path.
E. The marlin crossed the International Date Line along its path.

Have you ever visited a country where the law states that you drive on the opposite side of the road that you are accustomed to driving? It can provide some nervous moments the first time you enter a roundabout. When one tours this region you will not find a universal approach to traffic management in terms of left-hand or right-hand side driving. Open the *left or right?* folder and visit the four locations that are placemarked. Analyze the traffic patterns to determine what the local laws dictate.

5. Identify the correct pairing of Australia and Oceania locations and the side of the road on which traffic moves forward.

A. Australia – right; New Zealand – right; Samoa – left; Tahiti – left
B. Australia – left; New Zealand – left; Samoa – left; Tahiti – right
C. Australia – left; New Zealand – left; Samoa – right; Tahiti – right
D. Australia – right; New Zealand – right; Samoa – right; Tahiti – right
E. Australia – right; New Zealand – left; Samoa – left; Tahiti – right

Exploration 14.3: SHORT ESSAY

1. Explore the subculture of surfing. What are some of the unique attributes of surf culture? What is and is not valued?

____
____
____

2. Think about left-hand or right-hand driving patterns in regions like Australia and Oceania. What factors have contributed most potently to the distribution of traffic laws? Be sure to discuss Samoa's interesting recent history with this cultural attribute.

____
____
____

Exploration 14.4: AUSTRALIA & OCEANIA GEOPOLITICS

Political entities have a variety of different types of boundaries. Boundaries have not always been clearly delineated like they are in modern times. Just because a boundary is clearly defined does not mean that there will not be disputes over this delineation. Boundaries can be defined by natural features like rivers or mountain ranges (natural boundaries) or ethnographic features like religious or language divides (ethnographic boundaries). Geometric boundaries do not incorporate physical or cultural traits of the landscape. Rather they follow straight paths across the landscape. Boundaries are often composite types, meaning they incorporate two or more physical or cultural attributes. Expand the *Borders and Labels* folder in the *Primary Database* and then expand the *Borders* folder. Verify that the *1st Level Admin Borders* layer is turned on. Expand the GEOPOLITICS folder and then expand the *Boundaries* folder. View the borders of New Zealand, Australia, and Papua New Guinea and assess what type(s) of boundaries are utilized by each country.

1. Identify the statement that is most strongly supported by your assessment of state/province boundaries.

A. Australia uses ethnographic boundaries exclusively.
B. New Zealand only utilizes natural boundaries.
C. Papua New Guinea does not utilize geometric boundaries.
D. Australia and New Zealand use geometric boundaries while Papua New Guinea uses ethnographic boundaries.
E. All three states have evidence of using geometric boundaries to some extent.

2. Which of the following features is utilized as a natural boundary between states or provinces?

A. Rakaia River
B. Great Sandy Desert
C. Southern Alps
D. Great Dividing Range
E. Great Barrier Reef

NIMBY is an acronym that stands for not in my backyard and describes a situation where residents stand in opposition to a potential development or activity in close proximity to their homes. The term is usually applied to local contexts. For example, a neighborhood may stand in opposition to a new dump being created nearby or perhaps they do not want a new halfway house for recently released convicts. France and the United States applied the concept on a much broader scale in the South Pacific through the mid-1990s. Open the NIMBY folder and tour the four placemarks. It is difficult to see what occurred at these locations from the imagery alone. Only the *Ivy Mike* placemark gives a clear visible clue. Use the placemark names to learn more about the activities that occurred at these locations.

3. Based upon your analysis of the locations in the NIMBY folder combined with information from outside resources, which of the following statements is *least* accurate?

A. France tested weapons in the region until 1996.
B. Nuclear fallout poisoned many Pacific islands that were not evacuated for the tests.
C. The Castle Bravo shot was the largest of the nuclear tests.
D. Only underground nuclear tests took place at the Pacific test sites.
E. US nuclear testing took place at the sites collectively known as the Pacific Proving Grounds.

4. Examine Bikini Atoll, site of the Castle Bravo shot, and research the site using outside resources to evaluate which of the following statements is *least* accurate.

A. Bikini Atoll has been resettled and is currently home to several thousand islanders.
B. An air strip and small port provide access to Bikini Atoll.
C. The rough shape of the atoll is a result of coral activity rather than nuclear tests.
D. More than 20 nuclear weapons were tested at this site.
E. Originally, the islanders were relocated to Rongerik Atoll, before some returned in the 1970's.

Some islands in the region still seek to gain their independence. For example, the island of Bougainville is part of Papua New Guinea. Bougainville provides a disproportionate portion of Papua New Guinea's revenue because of a certain primary industry that takes place on the island. Twice the people of Bougainville sought independence from Papua New Guinea and they have since been granted a degree of autonomy from the rest of the country. Open the *independence?* placemark and survey the island. The most conspicuous anthropogenic feature of the island is the economic engine.

5. Bougainville generates most of its revenue via

   A. hunting whales.
   B. timber plantations.
   C. copper mining.
   D. oil and natural gas refining.
   E. ship manufacturing.

Exploration 14.4: SHORT ESSAY

1. Is the state or province where you live delineated by natural, ethnographic, or geometric boundaries? Research the history of your state or province boundaries and explain why your political entity has the shape that it has.

   ____
   ____
   ____

2. Open the independence here?! Placemark. What does the name of this placemark reference? Is there a Hawaiian independence movement? If so, what are the reasons behind this political action?

   ____
   ____
   ____

Exploration 14.5: AUSTRALIA & OCEANIA ECONOMY AND DEVELOPMENT

Extraction of heavy metal resources represents a large part of economies in the region. On the island of Papua, a similar feature can be seen as to what we viewed on Bougainville. Open the ECONOMY AND DEVELOPMENT folder and then the *OK Tedi Mine, Papua New Guinea* folder. Contained in the folder are image layers from 1990 and 2004. This mine is located in the high mountains of Papua New Guinea. The mine has had a staggering environmental impact on the region since it opened in 1984.

Exploration 14.5: MULTIPLE CHOICE

1. A big part of the environmental problems associated with the OK Tedi Mine stem from its position that is high in the OK Tedi River watershed. This high relief is demonstrated by the fact that the support town for the mine is less than 15 kilometers from the mine, but is approximately

   A. 1200 meters above the mine.
   B. 1200 meters below the mine.
   C. 5000 meters above the mine.
   D. 5000 meters below the mine
   E. 12,000 meters below the mine.

The adverse environmental impacts of the mine include problems associated with the more than 70 million tons of waste rock that are discharged from the mine annually. This discharge causes river beds to rise, which then leads to flooding, sediment deposition, and forest damage. The area's biodiversity has declined significantly according to the United Nations Environmental Program.

2. In terms of sedimentation from the OK Tedi Mine, what *cannot* be verified from the 1990 and 2004 image layers alone?

A. Sediment has dramatically affected at least two streams that flow from the mine area.
B. The stream channel associated with the more southern stream is more than five times its 1990 width by 2004.
C. Significant sediment accumulation can be seen downstream far beyond the extent of the image pair.
D. This area cannot be described as pristine as a result of the mine's activities.
E. No measures have been implemented to control the release of sediment from this mine.

A resource that is inherent to the region that can be a renewable resource if managed and harvested in a sustainable fashion are the numerous fisheries of the region. Of course many fisheries have not been managed responsibly and there are a number of aquatic species that have become threatened, endangered, or extinct. Threats beyond overfishing can include pollution and military activities such as sonar testing. In the *Primary Database,* expand the *Ocean* folder, and then turn on the *State of the Ocean* folder. Go to each placemark, and then click the nearby icon to learn about the status of several species in the region.

3. Identify the statement that is not supported by the information from the Monterey Bay Aquarium and the Marine Conservation Society.

A. Albacore tuna are the best choice for ocean-friendly tuna consumers.
B. Western Australia rock lobster should be avoided because it is not sustainably produced.
C. Bluefish tuna are severely overfished in all oceans.
D. Elevated levels of mercury have been found in bluefish tuna and longline-caught albacore tuna.
E. The best choice for healthy and sustainably produced crab is the Dungeness crab.

Turn off the *Ocean* folder and then dive into the 360° panoramas *overwater bungalow* and *moonrise*. These views give you a sampling of one of the region's primary economic drivers, tourism. The region has numerous resorts such as this one highlighted here that cater to the world's most affluent travelers. Check out the link to the resort website associated with the *island resort information* placemark.

4. Based on the panoramas, the Google imagery and the hotel website, which of the following statements is inaccurate?

A. The resort has cut/blasted out paths through the coral in the lagoon adjacent to the hotel property.
B. A room can cost in excess of 50,000 French Pacific Francs a night.
C. This appears to be the only resort on Moorea.
D. Moorea is a high island.
E. The overwater bungalows are not typical housing for Mooreans.

Tourism development has also boomed along a stretch of Australia known as the Gold Coast. Click the Gold Coast City, Queensland placemark to view the location. Verify that the 3D Buildings layer is turned on in the Primary Database and you will see one of the world's tallest residential buildings, Q1. A number of the residential units represent vacation homes for affluent people that have made their money by the export of iron ore to a trade partner with a growing need for this commodity. This trade partner has strengthened its trade relationship in terms of exports and imports at a steady pace over the last decade. Do some independent research to determine the country. Don't forget the iron ore hint.

5. With which of the following countries does Australia have the most rapidly growing trade partnership?

   A. China
   B. Japan
   C. New Zealand
   D. Samoa
   E. United States

Exploration 14.5: Short Answer

1. Were you familiar with the Monterey Bay Aquarium Seafood Watch before you completed this exercise? Open the link associated with the seafood 1 placemark and examine the recommendations and pocket guides for eating responsibly and safely when consuming seafood. How do some of your favorite items rank? Are they sustainable and/or do they have health risks?

   ____
   ____
   ____

2. Based simply on your evaluation of the overwater bungalow and moonrise bungalow along with you assessment of the surrounding landscape in Google Earth™, explain why you would or would not want to visit this location for a vacation. What attributes or characteristics are present (or not present) that make the location appealing or unappealing?

   ____
   ____
   ____

YOU MAP IT! Create Your Own KML Files.

In the *YOU MAP IT!* portion of this exploration, you will learn how to create a kml file with extruded polygons. This can be done in geographic information system software or with a number of freely available online tools. One of the best is kmlfactbook.org. It's very easy to use and employs contemporary data from the CIA World Factbook. Go to www.kmlfactbook.org. On the left-hand side of the page you will see a number of categories such as people, economy, and transport. You can expand these categories to see a wide variety of mappable variables. Select a variable and then click the "Preview in Map" button. Dive into this data with the intent of locating two data layers that illustrate the differences of the Australia and Oceania region from the rest of the world. When you have identified your first data set then click the "Download KML file" button. Click "Open" and you will see the layer loaded into your *Temporary Places* folder in Google Earth™. To keep this layer you will need to drag the kmlfactbook.org folder upward and into the *You Map It!* folder. Repeat this process for your second variable.

When you complete the Google Earth™ portion of this assignment, create a document that briefly interprets the patterns you see in your two kmls. Explain why you chose to map the variables you did and why they best represent regional differentiation for Australia and Oceania.

Turning it in:

Your instructor will provide you with an explanation of how to submit your results from this assignment. This may include e-mailing the .kml files you create from your *YOU MAP IT!* folder and a document with the required information to your instructor.

To create a .kmz from your *YOU MAP IT!* folder, simply click once on the *YOU MAP IT!* folder to highlight it, then go to File, Save Place As…, and save it in an appropriate location on your computer.